Pitman Research Notes in Mathematics Series

Submission of proposals for consideration

Suggestions for publication, in the form of outlines and representative samples, are invited by the Editorial Board for assessment. Intending authors should approach one of the main editors or another member of the Editorial Board, citing the relevant AMS subject classifications. Alternatively, outlines may be sent directly to the publisher's offices. Refereeing is by members of the board and other mathematical authorities in the topic concerned, throughout the world.

Preparation of accepted manuscripts

On acceptance of a proposal, the publisher will supply full instructions for the preparation of manuscripts in a form suitable for direct photo-lithographic reproduction. Specially printed grid sheets are provided and a contribution is offered by the publisher towards the cost of typing. Word processor output, subject to the publisher's approval, is also acceptable.

Illustrations should be prepared by the authors, ready for direct reproduction without further improvement. The use of hand-drawn symbols should be avoided wherever possible, in order to maintain maximum clarity of the text.

The publisher will be pleased to give any guidance necessary during the preparation of a typescript, and will be happy to answer any queries.

Important note

In order to avoid later retyping, intending authors are strongly urged not to begin final preparation of a typescript before receiving the publisher's guidelines and special paper. In this way it is hoped to preserve the uniform appearance of the series.

Longman Scientific & Technical
Churchill Livingstone Inc.
1560 Broadway
New York, NY 10036, USA
(tel (212) 819-5453)

Longman Scientific & Technical
Longman House.
Burnt Mill
Harlow, Essex, UK
(tel (0279) 26721)

Titles in this series

52 Wave propagation in viscoelastic media
F Mainardi
53 Nonlinear partial differential equations and their applications: Collège de France Seminar. Volume I
H Brezis and J L Lions
54 Geometry of Coxeter groups
H Hiller
55 Cusps of Gauss mappings
T Banchoff, T Gaffney and C McCrory
56 An approach to algebraic K-theory
A J Berrick
57 Convex analysis and optimization
J-P Aubin and R B Vintner
58 Convex analysis with applications in the differentiation of convex functions
J R Giles
59 Weak and variational methods for moving boundary problems
C M Elliott and J R Ockendon
60 Nonlinear partial differential equations and their applications: Collège de France Seminar. Volume II
H Brezis and J L Lions
61 Singular systems of differential equations II
S L Campbell
62 Rates of convergence in the central limit theorem
Peter Hall
63 Solution of differential equations by means of one-parameter groups
J M Hill
64 Hankel operators on Hilbert space
S C Power
65 Schrödinger-type operators with continuous spectra
M S P Eastham and H Kalf
66 Recent applications of generalized inverses
S L Campbell
67 Riesz and Fredholm theory in Banach algebra
B A Barnes, G J Murphy, M R F Smyth and T T West
68 Evolution equations and their applications
F Kappel and W Schappacher
69 Generalized solutions of Hamilton-Jacobi equations
P L Lions
70 Nonlinear partial differential equations and their applications: Collège de France Seminar. Volume III
H Brezis and J L Lions
71 Spectral theory and wave operators for the Schrödinger equation
A M Berthier
72 Approximation of Hilbert space operators I
D A Herrero
73 Vector valued Nevanlinna Theory
H J W Ziegler
74 Instability, nonexistence and weighted energy methods in fluid dynamics and related theories
B Straughan
75 Local bifurcation and symmetry
A Vanderbauwhede
76 Clifford analysis
F Brackx, R Delanghe and F Sommen
77 Nonlinear equivalence, reduction of PDEs to ODEs and fast convergent numerical methods
E E Rosinger
78 Free boundary problems, theory and applications. Volume I
A Fasano and M Primicerio
79 Free boundary problems, theory and applications. Volume II
A Fasano and M Primicerio
80 Symplectic geometry
A Crumeyrolle and J Grifone
81 An algorithmic analysis of a communication model with retransmission of flawed messages
D M Lucantoni
82 Geometric games and their applications
W H Ruckle
83 Additive groups of rings
S Feigelstock
84 Nonlinear partial differential equations and their applications: Collège de France Seminar. Volume IV
H Brezis and J L Lions
85 Multiplicative functionals on topological algebras
T Husain
86 Hamilton-Jacobi equations in Hilbert spaces
V Barbu and G Da Prato
87 Harmonic maps with symmetry, harmonic morphisms and deformations of metrics
P Baird
88 Similarity solutions of nonlinear partial differential equations
L Dresner
89 Contributions to nonlinear partial differential equations
C Bardos, A Damlamian, J I Díaz and J Hernández
90 Banach and Hilbert spaces of vector-valued functions
J Burbea and P Masani
91 Control and observation of neutral systems
D Salamon
92 Banach bundles, Banach modules and automorphisms of C*-algebras
M J Dupré and R M Gillette
93 Nonlinear partial differential equations and their applications: Collège de France Seminar. Volume V
H Brezis and J L Lions
94 Computer algebra in applied mathematics: an introduction to MACSYMA
R H Rand
95 Advances in nonlinear waves. Volume I
L Debnath
96 FC-groups
M J Tomkinson
97 Topics in relaxation and ellipsoidal methods
M Akgül
98 Analogue of the group algebra for topological semigroups
H Dzinotyiweyi
99 Stochastic functional differential equations
S E A Mohammed

100 Optimal control of variational inequalities
V Barbu
101 Partial differential equations and dynamical systems
W E Fitzgibbon III
102 Approximation of Hilbert space operators. Volume II
C Apostol, L A Fialkow, D A Herrero and D Voiculescu
103 Nondiscrete induction and iterative processes
V Ptak and F-A Potra
104 Analytic functions – growth aspects
O P Juneja and G P Kapoor
105 Theory of Tikhonov regularization for Fredholm equations of the first kind
C W Groetsch
106 Nonlinear partial differential equations and free boundaries. Volume I
J I Díaz
107 Tight and taut immersions of manifolds
T E Cecil and P J Ryan
108 A layering method for viscous, incompressible L_p flows occupying R^n
A Douglis and E B Fabes
109 Nonlinear partial differential equations and their applications: Collège de France Seminar. Volume VI
H Brezis and J L Lions
110 Finite generalized quadrangles
S E Payne and J A Thas
111 Advances in nonlinear waves. Volume II
L Debnath
112 Topics in several complex variables
E Ramírez de Arellano and D Sundararaman
113 Differential equations, flow invariance and applications
N H Pavel
114 Geometrical combinatorics
F C Holroyd and R J Wilson
115 Generators of strongly continuous semigroups
J A van Casteren
116 Growth of algebras and Gelfand–Kirillov dimension
G R Krause and T H Lenagan
117 Theory of bases and cones
P K Kamthan and M Gupta
118 Linear groups and permutations
A R Camina and E A Whelan
119 General Wiener–Hopf factorization methods
F-O Speck
120 Free boundary problems: applications and theory, Volume III
A Bossavit, A Damlamian and M Fremond
121 Free boundary problems: applications and theory, Volume IV
A Bossavit, A Damlamian and M Fremond
122 Nonlinear partial differential equations and their applications: Collège de France Seminar. Volume VII
H Brezis and J L Lions
123 Geometric methods in operator algebras
H Araki and E G Effros
124 Infinite dimensional analysis–stochastic processes
S Albeverio
125 Ennio de Giorgi Colloquium
P Krée
126 Almost-periodic functions in abstract spaces
S Zaidman
127 Nonlinear variational problems
A Marino, L Modica, S Spagnolo and M Degiovanni
128 Second-order systems of partial differential equations in the plane
L K Hua, W Lin and C-Q Wu
129 Asymptotics of high-order ordinary differentia equations
R B Paris and A D Wood
130 Stochastic differential equations
R Wu
131 Differential geometry
L A Cordero
132 Nonlinear differential equations
J K Hale and P Martinez-Amores
133 Approximation theory and applications
S P Singh
134 Near-rings and their links with groups
J D P Meldrum
135 Estimating eigenvalues with *a posteriori/a prio* inequalities
J R Kuttler and V G Sigillito
136 Regular semigroups as extensions
F J Pastijn and M Petrich
137 Representations of rank one Lie groups
D H Collingwood
138 Fractional calculus
G F Roach and A C McBride
139 Hamilton's principle in continuum mechanics
A Bedford
140 Numerical analysis
D F Griffiths and G A Watson
141 Semigroups, theory and applications. Volume
H Brezis, M G Crandall and F Kappel
142 Distribution theorems of L-functions
D Joyner
143 Recent developments in structured continua
D De Kee and P Kaloni
144 Functional analysis and two-point differential operators
J Locker
145 Numerical methods for partial differential equations
S I Hariharan and T H Moulden
146 Completely bounded maps and dilations
V I Paulsen
147 Harmonic analysis on the Heisenberg nilpote Lie group
W Schempp
148 Contributions to modern calculus of variation
L Cesari
149 Nonlinear parabolic equations: qualitative properties of solutions
L Boccardo and A Tesei
150 From local times to global geometry, control physics
K D Elworthy

151 A stochastic maximum principle for optimal control of diffusions
U G Haussmann
152 Semigroups, theory and applications. Volume II
H Brezis, M G Crandall and F Kappel
153 A general theory of integration in function spaces
P Muldowney
154 Oakland Conference on partial differential equations and applied mathematics
L R Bragg and J W Dettman
155 Contributions to nonlinear partial differential equations. Volume II
J I Díaz and P L Lions
156 Semigroups of linear operators: an introduction
A C McBride
157 Ordinary and partial differential equations
B D Sleeman and R J Jarvis
158 Hyperbolic equations
F Colombini and M K V Murthy
159 Linear topologies on a ring: an overview
J S Golan

L Boccardo & A Tesei (Editors)

University of Rome 'La Sapienza'/Second University of Rome

Nonlinear parabolic equations: qualitative properties of solutions

Longman Scientific & Technical

Copublished in the United States with
John Wiley & Sons, Inc., New York

Longman Scientific & Technical
Longman Group UK Limited
Longman House, Burnt Mill, Harlow
Essex CM20 2JE, England
and Associated Companies throughout the world.

Copublished in the United States with
John Wiley & Sons, Inc., 605 Third Avenue, New York, NY 10158

First published 1987

AMS Subject Classifications: 35K55, 35K65, 35R35, 49B22, 92A15.

ISSN 0269–3674

British Library Cataloguing in Publication Data
Nonlinear parabolic equations: qualitative
properties of solutions.—(Pitman
research notes in mathematics, ISSN 0269-
3674; 149).
1. Differential equations, Parabolic
2. Differential equations, Non Linear
I. Boccardo, L. II. Tesei, A.
515.3′53 QA374
ISBN 0-582-99459-4

Library of Congress Cataloging-in-Publication Data
Nonlinear parabolic equations.
(Pitman research notes in mathematics series,
ISSN 0269-3674; 149)
Proceedings of a conference held at the Second Rome
University from Apr. 1–5, 1985.
Includes bibliographies.
1. Differential equations, Parabolic. I. Boccardo,
L. (Lucio) II. Tesei, A. (Alberto) III. Series:
Pitman research notes in mathematics ; 149.
QA374.N59 1986 515.3′53 86-15699
ISBN 0-470-20379-X (USA only)

Printed and bound in Great Britain by
Biddles Ltd, Guildford and King's Lynn

Contents

Preface

This volume contains the proceedings of the conference "Nonlinear Parabolic Equations: Qualitative Properties of Solutions" held at the Second Rome University from April 1 to April 5, 1985.

Nonlinear diffusion problems arising in the applied sciences have been widely investigated in recent years. They exhibited stimulating connections with well understood, more "abstract" subjects in the theory of nonlinear parabolic equations (like comparison principles, or existence, uniqueness and regularity of solutions). At the same time, seeking solutions in some special class motivated problems in ODE's or elliptic PDE's.

The reader will trace himself such links through the different topics of the conference:

(a) existence, uniqueness and regularity of solutions for parabolic equations and systems;
(b) special classes of solutions;
(c) free boundary problems;
(d) degenerate diffusion;
(e) asymptotical behaviour of solutions;
(f) models from biology, physics and engineering.

We hope that these proceedings bear evidence of the interest and the rapid evolution of this field.

We are glad to acknowledge support from the following institutions, that made the conference possible:

- SECONDA UNIVERSITA' DI ROMA
- CONSIGLIO NAZIONALE DELLE RICERCHE (Comitato Nazionale per le Scienze Matematiche and Gruppo Nazionale per l'Analisi Funzionale e Applicazioni)
- MINISTERO DELLA PUBBLICA ISTRUZIONE (Gruppo Nazionale "Equazioni Differenziali").

We are particularly grateful to the Rector of the Second Rome University, Enrico Garaci, for his warm and helpful interest. We also wish to thank Roberto Conti, Piero de Mottoni and Carlo D. Pagani for their concern and useful suggestions.

Special thanks are due to our colleagues in the organizing committee, Daniela Giachetti and Maria Assunta Pozio, whose valuable contribution was so effective for the success of the meeting.

Last but not least, let us express our gratitude to all the participants: their presence made the conference a rewarding experience, not only from the scientific point of view.

Rome, January 1986

Lucio Boccardo

Alberto Tesei

List of contributors

H. Amann: Mathematisches Institut, Universität Zürich, CH-8001 Zürich, Switzerland.

F.V. Atkinson: Department of Mathematics, University of Toronto, Canada.

C. Bandle: Mathematisches Institut, Universität Basel, CH-4051 Basel, Switzerland.

A. Bensoussan: INRIA, 78153 Le Chesnay, France.

M. Biroli: Dipartimento di Matematica, Politecnico di Milano, 20133 Milano, Italia.

G. Caginalp: Mathematics Department, University of Pittsburgh, PA 15260, USA.

S. Campanato: Dipartimento di Matematica, Università di Pisa, 56100 Pisa, Italia.

F. Chiarenza: Dipartimento di Matematica, 95125 Catania, Italia.

G. Da Prato: Scuola Normale Superiore, 56100 Pisa, Italia.

R. Dal Passo: Istituto per le Applicazioni del Calcolo, 00161 Roma, Italia.

E. Di Benedetto: Department of Mathematics, Northwestern University IL 60202, USA.

J.I. Diaz: Departamento de Ecuaciones Funcionales, Facultad de Matemáticas, Universidad Complutense, 28040 Madrid, Espana.

A. Fasano: Istituto Matematico 'U. Dini', Università di Firenze, 50134 Firenze, Italia.

P.C. Fife: Mathematics Department, University of Arizona, AZ 85721, USA.

A. Friedman: Department of Mathematics, Northwestern University, IL 60201, USA.

M.G. Garroni: Dipartimento di Matematica, Università di Roma 'La Sapienza', 00185 Roma, Italia.

J. Hernandez: Departamento de Matemáticas, Universidad Autónoma, 28049 Madrid, Espana.

M.A. Herrero: Departamento de Ecuaciones Funcionales, Facultad de Matemáticas, Universidad Complutense, 28040 Madrid, Espana.

P.Hess: Mathematisches Institut, Universität Zürich, CH-8001 Zürich, Switzerland.

S. Kamin: School of Mathematical Sciences, Tel-Aviv University, Tel-Aviv, Israel.

R. Kersner: Magyar Tudományos Akadémia, Számítastechnikai és Automatizálási Kutató Intézete, H-1502 Budapest, Hungary.

S. Luckhaus: Sonderforschungsbereich 123, 6900 Heidelberg, BRD.

A. Lunardi: Dipartimento di Matematica, Università di Pisa, 56100 Pisa, Italia.

H. Matano: Department of Mathematics, Hiroshima University, Japan.

M. Mimura: Department of Mathematics, Hiroshima University, Japan.

T. Nagai: Department of Mathematics, Faculty of Education, Ehime University, Matsuyama, Japan.

Y. Nishiura: Institute of Computer Sciences, Kyoto Sangyo University, Kyoto 603, Japan.

L.A. Peletier: Rijksuniversiteit te Leiden, Subfaculteit der Wiskunde en Informatica, 2300 RA Leiden, Nederland.

M. Pierre: Département de Mathématiques, Université de Nancy, France.

J.-P. Puel: Laboratoire D'Analyse Numérique, Université Paris VI, Paris, France.

M. Schatzman: Mathématiques, Université Claude-Bernard, 69 622 Villeurbanne CEDEX, France.

E. Sinestrari: Dipartimento di Matematica, Università di Roma 'La Sapienza', 00185 Roma, Italia.

A. Tesei: Dipartimento di Matematica, Seconda Università di Roma, 00173 Roma, Italia.

G.M. Troianiello: Dipartimento di matematica, Università di Roma I, Italia.
A. Visintin: Istituto di Analisi Numerica, 27100 Pavia, Italia.

H AMANN

Quasilinear parabolic systems

Let Ω be a bounded domain in $\mathbb{R}^n$ with a smooth boundary $\partial\Omega$. Then we consider initial boundary value problems of the form

$$\begin{cases} \dfrac{\partial u}{\partial t} + A(t,u)u = f(x,t,u,\nabla u) & \text{in } \Omega \times (0,T], \\ \mathcal{B}(t,u)u = g(x,t,u) & \text{on } \partial\Omega \times (0,T], \\ u(\cdot,0) = u_o & \text{on } \Omega. \end{cases} \tag{1}$$

We assume that

$$A(t,u)u := -\partial_j(a_{jk}(\cdot,t,u)\partial_k u) + a_j(\cdot,t,u)\partial_j u,$$

where we use the summation convention, j,k running from 1 to n, and that

$$\mathcal{B}(t,u)u := a_{jk}(\ ,t,u)\nu^j\partial_k u,$$

where $\nu = (\nu^1,\ldots,\nu^n)$ denotes the outer normal on $\partial\Omega$. The coefficients of these operators are supposed to be smooth matrix-valued functions, that is,

$$a_{jk} = a_{kj}, a_j \in C^\infty(\bar{\Omega} \times [0,T] \times \mathbb{R}^N, \ \mathcal{L}(\mathbb{R}^N)), \ \ 1 \leq j,k \leq n,$$

where N is a positive integer and $\mathcal{L}(\mathbb{R}^N)$ the space of all endomorphisms of $\mathbb{R}^N$ (= the space of all N × N-matrices).

In addition we assume that A is a parabolic linear operator, where we allow a rather general meaning of the word "parabolic". For example, A is parabolic if it is strongly parabolic, that is, if

$$(a_{jk}(x,t,\eta)\zeta|\zeta)\xi^j\xi^k > 0$$

for all $(x,t,\eta) \in \bar{\Omega} \times [0,T] \times \mathbb{R}^N$, all $\xi = (\xi^1,\ldots,\xi^n) \in \mathbb{R}^n \setminus \{0\}$ and all $\zeta \in \mathbb{R}^n\setminus\{0\}$, where $(\cdot|\cdot)$ is the usual euclidean inner product in $\mathbb{R}^N$.

However our results apply also to systems which are not strongly parabolic. For example, A could be of the form

$$A(t,u) = \begin{bmatrix} A^{11}(t,u) & A^{12}(t,u) \\ & \\ 0 & A^{22}(t,u) \end{bmatrix},$$

where $A^{jj}(t,u)$, $j = 1,2$, are strongly parabolic operators acting on N_j-vector valued functions with $N_1 + N_2 = N$ and with no restriction for the operator $A^{12}(t,u)$. Systems of this form occur in many applications, and it is easily seen that they are not strongly parabolic, in general.

Finally we assume that

$$f : \overline{\Omega} \times [0,T] \times \mathbb{R}^N \times \mathbb{R}^{nN} \to \mathbb{R}^N$$

and

$$g : \partial\Omega \times [0,T] \times \mathbb{R}^N \to \mathbb{R}^N$$

are smooth functions such that

$$g(x,t,0) = 0, \qquad (x,t) \in \partial\Omega \times [0,T]. \tag{2}$$

By a <u>classical solution</u> u of (1) on an interval $J \subset [0,T]$ with $0 \in J$ and $\dot{J} := J\setminus\{0\} \neq \emptyset$, we mean a function

$$u \in C(\overline{\Omega} \times J, \mathbb{R}^N) \cap C^1(\overline{\Omega} \times \dot{J}, \mathbb{R}^N) \cap C^{2,0}(\Omega \times \dot{J}, \mathbb{R}^N),$$

which satisfies (1) pointwise, where $C^{2,0}$ is the set of all functions which are continuous and twice continuously differentiable with respect to the first variable.

In the following we denote by $W_p^\tau := W_p^\tau(\Omega, \mathbb{R}^N)$ the standard Sobolev-Slobodeckii spaces. Then we can formulate our main result, namely the following

THEOREM. *Suppose that* $1 + n/p < \tau \leqq 2$ *and that* $u_0 \in W_p^\tau$ *satisfies the compatibility condition*

$$\mathcal{B}(0,u_0)u_0 = g(\cdot,0,u_0) \qquad \text{on } \partial\Omega.$$

Then problem (1) *possesses a maximal classical solution* u *defined on an open subinterval* J *of* [0,T]. *If*

$$\sup_{t\in J} \|u(t)\|_{W_p^\tau} < \infty$$

then J = [0,T], *that is,* u *is a global solution. If* f *is independent of* ∇u *then* u *is the only solution of* (1).

The proof of this theorem is based upon general results concerning abstract evolution equations of the form

$$\dot{u} + A(t,u)u = F(u) \tag{3}$$

in general Banach spaces. Here -A(t,u) is for each fixed argument the infinitesimal generator of a strongly continuous analytic semigroup. The main difficulty stems from the fact that the domains D(A(t,u)) vary with (t,u), in general. To overcome this difficulty problem (3) is extended by an appropriate "extrapolation method" to a superspace so that the extended linear operators have constant domain. Then sharp estimates in interpolation spaces for linear evolution equations with time-dependent principal part having constant domain are used, in conjunction with fixed point arguments and regularity properties of parabolic fundamental solutions, to obtain the desired results.

Several remarks are in order:

(a) The abstract theorems apply also to some parabolic systems of higher order.

(b) It can be shown that the solution u of (1) depends continuously upon the initial value u_0 if f is independent of ∇u.

(c) There are no growth or compatibility conditions for the nonlinearities f and g besides (2).

(d) If the boundary operator $\mathcal{B}(t,u)$ is replaced by a boundary operator $\mathcal{B}$ which is independent of t and u - for example this is the case if $\mathcal{B}$ denotes the boundary operator induced by Dirichlet boundary conditions

- and if $g = 0$, then one gets uniqueness also in the case that f depends upon ∇u. Moreover in this case the map sending the initial function u_o into the corresponding solution of (1) defines a (local) semiflow on the Banach space $\{u \in W_p^\tau ; \mathcal{B}u = 0\}$, provided the system (1) is autonomous.

For further results and proofs we refer to the papers cited below.

REFERENCES

1. Amann, H.: Quasilinear evolution equations and parabolic systems, Trans. Amer. Math. Soc. (to appear).

2. Amann, H.: On abstract parabolic fundamental solutions (to appear).

3. Amann, H.: Quasilinear parabolic systems under nonlinear boundary conditions (to appear).

H. Amann
Mathematisches Institut
Universität Zürich
Rämistrasse 74
CH-8001 Zürich
Switzerland

F V ATKINSON & L A PELETIER

Radial similarity of a parabolic equation

We complete and extend, particularly for the case $N \in (2,4)$, certain results concerning the non-existence of rapidly decreasing positive solutions of the Haraux-Weissler equation

$$u'' + ((N-1)/r + \tfrac{1}{2}r)u' + \tfrac{1}{2}\lambda u + |u|^{p-1}u = 0, \tag{1}$$

previously considered [1] by the authors with the aid of special functions. The method is generalized and applied to a slightly more general equation than (1).

1. Introduction

The equation (1) arises in the search for solutions of the non-linear heat equation

$$w_t = \Delta w + t^m w^p \quad (m > -1,\ p > 1), \tag{2}$$

in the form

$$w(t,x) = t^{-n}G(rt^{-\frac{1}{2}}); \tag{3}$$

here $t > 0$, $x \in R^N$, $N > 2$, $r = |x|$ (see e.g. [3], [4], [5]). Taking $n = (m+1)/(p-1)$, one is led to (1) with $\tfrac{1}{2}\lambda = n$. There has been particular interest lately in the study of the solution of (1) such that

$$u(0) = \gamma > 0,\ u'(0) = 0, \tag{4}$$

more especially in its behaviour as $r \to \infty$.

It is known ([4], [5]) that just three types of asymptotic behaviour are possible, namely:

(i) positive and slowly decreasing, with

$$\lim r^{\lambda}u(r) \in (0,\infty), \tag{5}$$

as $r \to \infty$;

(ii) positive and rapidly decreasing, with

$$u(r) \sim A \exp(-\tfrac{1}{4}r^2) r^{\lambda-N}, \qquad (6)$$

for some $A \in (0,\infty)$;

(iii) u fails to be positive on $(0,\infty)$.

The problem is to determine, so far as possible, how this behaviour depends on the choice of the four parameters N, λ, p and γ.

We may term these solution-classes "slow", "fast" and "zero-type", respectively, indicating their occurrence on diagrams by the letters S, F and Z.

Similar questions can be posed for a slightly more general equation. If $u(r)$ is interpreted as a spherically symmetric function on R^N, namely $u(x)$ with $x \in R^N$, then (1) can be re-written in the form

$$\Delta u + \tfrac{1}{2}x.\nabla u + f(u) = 0,$$

or

$$K^{-1}\Delta(Ku) + f(u) = 0, \quad K = \exp(\tfrac{1}{4}r^2), \qquad (7)$$

which, for a general class of K, has been studied by Escobedo and Kavian [2]. This suggests the study of

$$u'' + \varphi(r)u' + \tfrac{1}{2}\lambda u + |u|^{p-1} u = 0, \qquad (8)$$

where

$$\varphi(r) = (N-1)/r + \eta(r), \qquad (9)$$

and $\eta(r)$ is some suitably restricted function.

It will be assumed throughout that $N > 2$. Methods and motivation from PDE sources are naturally confined to the case of integral N; however we do not need that N should be an integer.

We shall be mainly interested in the "supercritical" case, in which

$$p \geq (N+2)/(N-2). \qquad (10)$$

We review three types of result on the classification of the parameter-values in respect of the asymptotic behaviour of $u(r)$, presenting them in order of immediacy.

2. The first result

The following is to be seen by simple Sturmian arguments, and is known (see e.g. [4], [5]).

THEOREM 1. *For the equation* (1) *we have*

(a) $\lambda > N \Rightarrow Z$;

(b) $0 < \lambda < N$, $\gamma^{p-1} < \frac{1}{2}(N-\lambda) \Rightarrow S$.

Here "Z", "S" refer to the classifications (iii), (i) of Section 1. In the (λ,γ)-plane this may be represented thus:

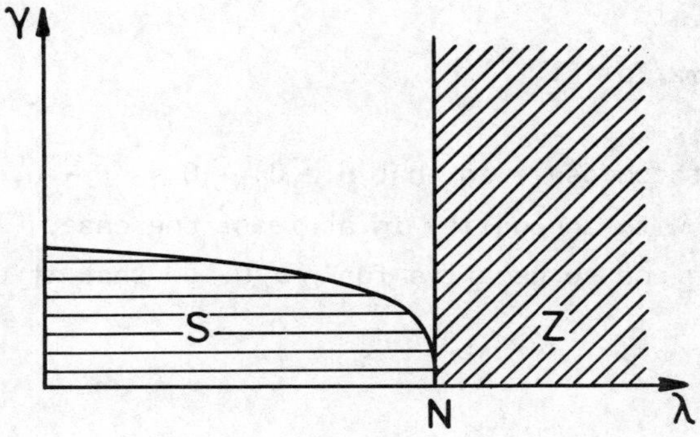

Fig. 1. $p > 1$.

This may be seen somewhat more generally. In (9) we take it that

$$\eta \in C''[0,\infty),\ \eta'(0) \geq \tfrac{1}{2},\ \eta''(r) \geq 0, \tag{11}$$

and write

$$\zeta(r) = \int_0^r \eta(s)ds,\ \psi = r^{N-1} \exp \zeta,\ y = r/\psi. \tag{12}$$

It is then seen that $y(r)$ satisfies

$$y'' + \varphi y' + (\eta' + (N-1)\eta/r)y = 0. \tag{13}$$

Comparison arguments between (8) and (13) then lead to conclusion (b) of the theorem, and to (a) in the modified form that we should have $\lambda > 2N \sup \eta'(r)$ with the latter to be finite.

3. A "Pohožaev identity"

For more delicate results one introduces the functional

$$H(z,u) = z\psi^2 u'^2 - z'\psi^2 uu' + 2z\psi^2 F(u) + \tfrac{1}{4}(z'\psi)'\psi u^2, \tag{14}$$

where, as usual, $F(u) = \int_0^u f(v)dv$, and z is some suitably smooth function. For a solution u of (8) one finds that, with $H' = dH/dr$,

$$H'(z,u) = \tfrac{1}{2}u^2\psi^2 P(z) + |u|^{p+1}\psi^2 Q(z), \tag{15}$$

where

$$P(z) = z''' + 3\varphi z'' + (\varphi' + 2\varphi^2 + 2\lambda)z' + 2\lambda\varphi z \tag{16}$$

and

$$Q(z) = z' + 4\varphi z/(p+1). \tag{17}$$

The aim is then to choose z so that $H(z,u) \to 0$ as $r \to 0$, and so that also $P(z) \leq 0$, $Q(z) \leq 0$, with inequality in at least one case. It then follows that H is decreasing and so negative for $r > 0$, so that it cannot vanish, nor tend to 0 as $r \to \infty$.

4. A second result

It may be checked that the choice $z(r) = y(r)$ (see (12)) satisfies the above requirements, so that one gets the following result, known for (1) (see [5]).

THEOREM 2. $0 < \lambda \leq \tfrac{1}{2}N$, $p \geq (N+2)/(N-2) \Rightarrow S$.

We comment that this also applies to (8), subject to (11).

Diagrammatically, one gets a slight amplification of Fig. 1, though for a more restricted range of p.

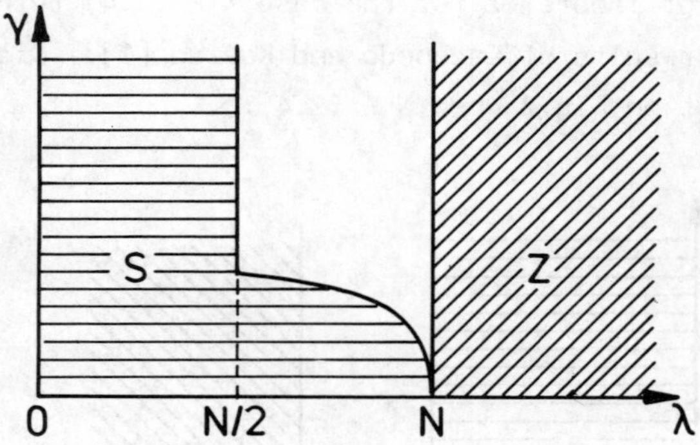

Fig. 2. $p \geq (N+2)/(N-2)$.

5. A third result

This depends on a more recondite choice of z in the above argument. This is based on the curious fact that solutions of the third-order differential equation $P(z) = 0$ with P as in (16) are given by products of two solutions of the linearized version of (8), that is to say

$$u'' + \varphi u' + \tfrac{1}{2}\lambda u = 0. \tag{18}$$

More precisely, if α,β are linearly independent solutions of (18), then the general solution of $P(z) = 0$ will be given by a linear combination of α^2, $\alpha\beta$, and β^2. We are here concerned with a choice $z = \alpha\beta$, and so will have $P(z) = 0$, and need that

$$Q(z) = Q(\alpha\beta) < 0.$$

We achieve this by requiring that

$$\alpha > 0,\ \beta > 0,\ \alpha'/\alpha + \beta'/\beta + 4\varphi/(p+3) < 0. \tag{19}$$

In addition, of course, the behaviour of α,β as $r \to 0$ must be such that $H(\alpha\beta,u) \to 0$.

By these means, for the case of (1), we proved in [1]:

THEOREM 3. $0 < \lambda < 2$, $p \geq (N+2)/(N-2) \Rightarrow S$.

This improvement of Theorem 2 for the case $N \in (2,4)$ turns out to dovetail with certain recent results of Escobedo and Kavian [2]. Diagrammatically, we get an improvement on Fig.2 when $N \in (2,4)$:

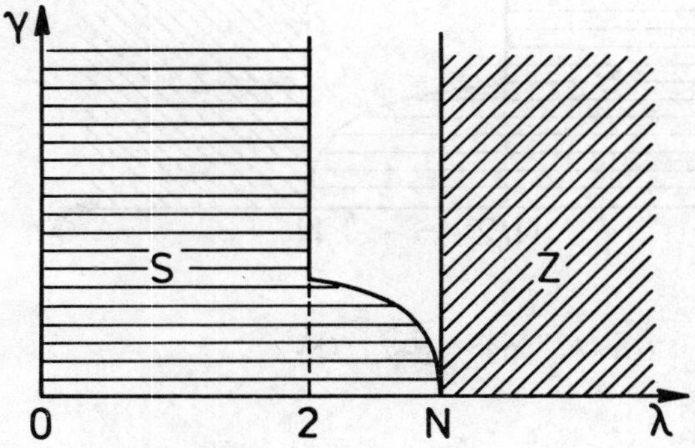

Fig. 3. $p \geq (N+2)/(N-2)$, $N \in (2,4)$.

The proof given in [1] for the case of (1) relied on the fact that solutions of the linearized equation (18) in this case, with $\varphi(r)=(N-1)/r+\frac{1}{2}r$, can be given explicitly in terms of special functions. Writing

$$g(r,t) = \exp\,(-\tfrac{1}{4}r^2t)t^{\frac{1}{2}\lambda-1}|1-t|^{\frac{1}{2}(N-\lambda)-1}$$

we took

$$\alpha(r) = \int_0^1 g(r,t)dt, \quad \beta(r) = \int_1^\infty g(r,t)dt.$$

However our remarks concerning the solution of $P(z) = 0$ in terms of the solutions of (18) hold good even when explicit solutions of (18) are not to hand. We must then rely on the methods of qualitative theory to establish the existence and requisite properties of $\alpha(r)$, $\beta(r)$.

6. Sketch of the more general case

For the case that α,β cannot be given explicitly, we define $\alpha(r)$ as the solution of (18) such that $\alpha(0) = 1$, $\alpha'(0) = 0$. Subject to validation we then set

$$\beta(r) = \alpha(r) \int_r^\infty \{\psi(s)\alpha^2(s)\}^{-1} ds, \tag{20}$$

where ψ is given in (12). We note the main steps in the argument, assuming (8), (9), (11) and bounds on λ as given.

LEMMA 1. *If* $0 < \lambda < 2N\eta'(0)$, *then* $\alpha(r) > y(r)$ *and* $\alpha'(r) < 0$ *for* $r > 0$.

Since $\alpha' \le 0$, the term α'/α in (19) may be dropped.

LEMMA 2. *If* $0 < \lambda < 2N\eta'(0)$, *the solution* $\beta(r)$ *given in* (20) *exists, is positive on* $(0,\infty)$, *and satisfies* $\beta(r) = O(y(r))$ *as* $r \to \infty$.

The proof of convergence in (20) requires an improved lower estimate for $\alpha(r)$ when r is large, of the form

$$\alpha(r) > \text{const. } y(r) \int_1^r \{\psi(s)y^2(s)\}^{-1} ds > 0.$$

Finally, we need

LEMMA 3. *If* $0 < \lambda < 4\eta'(0)$, *then*

$$\beta'/\beta + (N-2)/r + \eta(r) < 0. \tag{21}$$

This depends on Wronskian arguments involving β and y.

On substituting (21) in (19) we see that it is sufficient that

$$- (N-2)/r - \eta(r) + 4\varphi(r)/(p+3) < 0,$$

which leads again to $p \ge (N+2)/(N-2)$. Thus Theorem 3 may be extended to the (8), (9), (11), with the requirement that $0 < \lambda < 4\eta'(0)$.

REFERENCES

1. Atkinson, F.V. and L.A. Peletier: Sur les solutions radiales de l'equation $\Delta u + \frac{1}{2}x.\ \nabla u + \frac{1}{2}\lambda u + |u|^{p-1} u = 0$, C.R. Acad. Sci. Paris (to appear).
2. Escobedo, M. and O. Kavian: Variational problems related to self-similar solutions of the heat equation, (preprint).

3. Giga, Y. and R.V. Kohn: Asymptotically self-similar blow-up of semilinear heat equations (preprint).

4. Haraux, A. and F.B. Weissler: Non-uniqueness for a semilinear initial value problem, Indiana Univ. Math. J. 31 (1982), 167-189.

5. Peletier, L.A., D. Terman and F.B. Weissler: On the equation $\Delta u + \frac{1}{2}x.\nabla u + f(u) = 0$ (Report 84-21, University of Leiden).

F.V. Atkinson
Department of Mathematics
University of Toronto
Toronto, Ontario M5S 1A1
Canada

L.A. Peletier
Rijksuniversiteit te Leiden
Subfaculteit der Wiskunde en Informatica
Wassenaarseweg 80
Postbus 9512
2300 RA Leiden
Nederland

F V ATKINSON & L A PELETIER

Elliptic equations with critical exponents

1. Introduction

Consider the problem

$$\left.\begin{array}{llll} -\Delta u = u^q + u^p, & u > 0 & \text{in} & B_R \\ u = 0 & & \text{on} & \partial B_R \end{array}\right\} \quad (I) \qquad (1)$$

in which B_R is the open ball with radius R in $\mathbb{R}^N$ centred around the origin and

$$p = \frac{N+2}{N-2}, \quad 1 \leq q < p, \quad N > 2.$$

In this lecture we shall discuss the existence and uniqueness of solutions of Problem I for different values of the parameters R, q and N.

In a recent paper [4] Brezis and Nirenberg established the following results about Problem I.

<u>q = 1</u>. If $N \geq 4$, Problem I has a solution if and only if $0 < R < \lambda_o^{\frac{1}{2}}$, but if $N = 3$, it only has one if and only if $\frac{1}{2}\lambda_o^{\frac{1}{2}} < R < \lambda_o^{\frac{1}{2}}$. Here λ_o denotes the principal eigenvalue of $-\Delta$ on the unit ball.

<u>q > 1</u>. If $N \geq 4$, Problem I has a solution for every $R > 0$ but if $N = 3$ there are two cases:

(a) if $3 < q < 5$, there exists a solution for every $R > 0$;

(b) if $1 < q \leq 3$, there exists a solution if and only if R is sufficiently large.

In addition, they made the conjecture, based on numerical evidence, that, when $N = 3$ and $1 < q < 3$, there are <u>two</u> solutions when R is large enough.

The solutions referred to above all necessarily have radial symmetry [7]. Exploiting this property, we shall sketch in this lecture an ODE-proof of some of these results, prove the non-uniqueness conjecture and describe the dependence of the solution u on the parameters R, q and N, where the latter now need not be an integer. For further details we refer to [1], [2]. Special emphasis will be given to the rôle of the critical value N = 4.

2. Formulation

We look for radial solutions $u = u(r)$, where $r = |x|$. They are solutions of the initial value problem

$$\left.\begin{array}{l} u'' + \frac{N-1}{r} u' + f(u) = 0, \; u > 0 \; (0 < r < R) \\ u'(0) = 0, \quad u(R) = 0 \end{array}\right\} \quad (II) \tag{2}$$

where $f(u) = u^q + u^p$ and $1 \leq q < p = (N+2)/(N-2)$.

Equation (2) can be transformed to a generalized Emden-Fowler equation [5], [6] if we set

$$t = (N-2)^{N-2} r^{-(N-2)}, \quad y(t) = u(r). \tag{3}$$

We then obtain

$$\left.\begin{array}{l} y'' + t^{-k} f(y) = 0, \; y > 0 \; (T < t < \infty) \\ y(T) = 0, \quad y'(\infty) = 0 \end{array}\right\} \quad (III) \tag{4}$$

where

$$k = 2\,\frac{N-1}{N-2} \quad \text{and} \quad T = (N-2)\, R^{-(N-2)}. \tag{5}$$

Because $k > 2$ for every $N > 2$, there exists for every $\gamma > 0$ a unique solution $y(t,\gamma)$ of equation (4) with the property

$$\lim_{t\to\infty} y(t,\gamma) = \gamma. \tag{6}$$

Let $T = T(\gamma)$ denote the "first" zero of $y(t,\gamma)$ as t decreases from ∞ (Fig. 1):

$$T(\gamma) = \inf\,\{t > 0 : y(\cdot,\gamma) > 0 \;\; \text{on} \;\; (t,\infty)\}. \tag{7}$$

This zero corresponds to a zero $R(\gamma)$ of the solution $u(r,\gamma)$ of (2) with initial value $u(0,\gamma) = \gamma$, and initial slope $u'(0,\gamma) = 0$ (see Fig. 1).

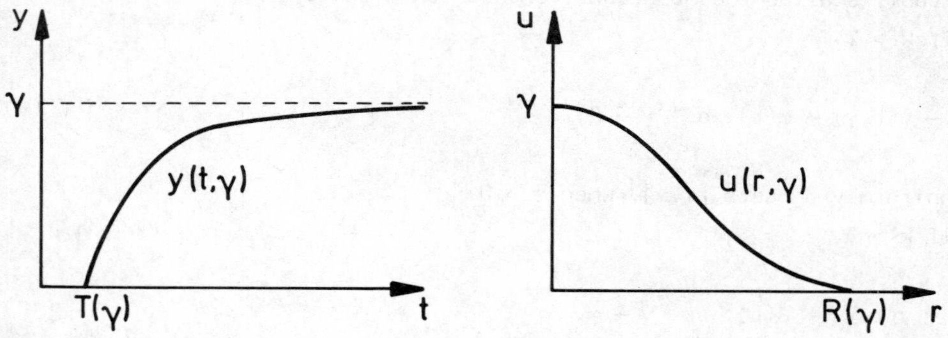

Fig. 1. The graphs of $y(t;\gamma)$ and $u(r,\gamma)$.

Clearly, for fixed values of q and N, the set of values of R for which Problems I and II have a solution, or the values of T for which Problem III has a solution, is completely described by the range of the functions $R(\gamma)$, respectively $T(\gamma)$, as γ runs from 0 to ∞.

REMARKS. 1. $(N+2)/(N-2) = 2k - 3$.

2. If $f(y) = y^{\sigma}$, $\sigma > 0$, then equation (4) becomes the well-known Emden-Fowler equation. If moreover, $\sigma = 2k - 3$, it has the Emden solution

$$y_o(t,\gamma) = \gamma t \left(t^{k-2} + \frac{1}{k-1}\gamma^{2k-4}\right)^{-1/(k-2)}.$$

3. One can prove the following upper bound for solutions $y(t,\gamma)$ of (4) and (6) [1].

LEMMA 1. *If* $k > 2$ *and* f *satisfies*

$$uf'(u) \leq (2k-3)f(u) \quad \textit{for all} \quad u > 0, \tag{8}$$

then

$$y(t,\gamma) < z(t,\gamma) \quad \textit{for all} \quad t \geq T(\gamma),$$

where

$$z(t,\gamma) = \gamma t \left\{t^{k-2} + \frac{1}{k-1}\gamma^{-1} f(\gamma)\right\}^{-1/(k-2)}.$$

Note that $f(u) = u^q + u^{2k-3}$ satisfies (8) if $1 \leq q < 2k - 3$.

3. The case $q = 1$

By suitably scaling the solution y one can show that:

(i) if $k > 2$

$$\frac{1}{\gamma}\, y(t,\gamma) \to \alpha(t) \quad \text{as} \quad \gamma \to 0, \tag{9}$$

uniformly on sets $[\tau,\gamma]$, when $\tau > 0$.

(ii) if $k > 3$

$$\gamma\, y(t,\gamma) \to \beta(t) \quad \text{as} \quad \gamma \to \infty, \tag{10}$$

uniformly on sets $[\tau,\gamma]$, when $\tau > 0$.

Here α and β are solutions of the equation

$$y'' + t^{-k}y = 0, \tag{11}$$

which exhibit the following behaviour at infinity:

$$\alpha(t) \to 1 \qquad \text{as} \quad t \to \infty \tag{12a}$$
$$\beta(t) - k_1 t \to 0 \qquad \text{as} \quad t \to \infty, \tag{12b}$$

where $k_1 = (k-1)^{1/(k-2)}$.

Let us denote the first zero of α by T_o and of β by T_1. We can establish the following properties of the function $T(\gamma)$.

<u>THEOREM 1.</u> *Suppose* $q = 1$ *and* $k > 2$.

(i) $T(\gamma) > T_o$ *for all* $\gamma > 0$

and

$$\lim_{\gamma\to\infty} T(\gamma) = T_o.$$

(ii) *If* $k \le 3$,

$$\lim_{\gamma\to\infty} T(\gamma) = \infty.$$

(iii) *If* $k > 3$, *then*

$$T(\gamma) < T_1 \quad \textit{for all} \quad \gamma > 0$$

and

$$\lim_{\gamma\to\infty} T(\gamma) = T_1.$$

Returning to the original variables, we see that the functions α and β are positive radial solutions of the problem

$$\Delta u + u = 0 \quad \text{in} \quad B_{R_i}$$

$$u = 0 \quad \text{on} \quad \partial B_{R_i}$$

where $R_i = (N-2)T_i^{-1/(N-2)}$, $i = 0,1$. Near the origin α is regular:

$$\alpha(r) \to 1 \quad \text{as} \quad r \to 0,$$

but β is singular:

$$\beta(r) \sim c(N)r^{-(N-2)} \quad \text{as} \quad r \to 0,$$

where $c(N) = (N(N-2))^{(N-2)/2}$.

If we translate the results of Theorem 1 to the r-variable, we obtain as $\gamma \to 0$:

$$R(\gamma) \to R_0 \text{ when } \quad N > 2$$

and as $\gamma \to \infty$:

$$R(\gamma) \to \begin{cases} R_1 & \text{when} \quad 2 < N < 4 \\ 0 & \text{when} \quad N \geq 4 \end{cases}.$$

4. The case $q > 1$

We shall derive estimates for $T(\gamma)$ as $\gamma \to 0$ and as $\gamma \to \infty$.

THEOREM 2. *Suppose* $q > 1$. *Then*

$$T(\gamma) = O(\gamma^{(q-1)/(k-2)}) \quad \textit{as } \gamma \to 0.$$

Proof. We integrate the equation for y twice, the second time over (T,∞), where $T = T(\gamma)$. Then

$$0 = \gamma - \int_T^\infty (s-T)s^{-k} f(y(s,\gamma))ds$$

$$\geq \gamma - \frac{f(\gamma)}{(k-1)(k-2)} T^{2-k},$$

from which our estimate follows.

To obtain a first estimate for $T(\gamma)$ as $\gamma \to \infty$, we recall from Lemma 1 the upper bound

$$y(t,\gamma) < z(t,\gamma).$$

Since $y(t,\gamma)$ satisfies the integral equation

$$y(t,\gamma) = \gamma - \int_t^\infty (s-t)s^{-k} f(y(s,\gamma))ds, \tag{13}$$

we obtain a lower bound for y if we replace y by z in the right-hand side of (13). This yields eventually, if $q > k-2$,

$$\gamma y(t,\gamma) > \frac{1}{1+\gamma^{3-2k+q}} \left(\gamma z(t,\gamma) - \frac{k-1}{q-k+2} \gamma^{5-2k+q} \right). \tag{14}$$

Observing that the zero of the right-hand side of (14) provides our upper bound for $T(\gamma)$ we obtain the following estimate.

LEMMA 2. *Suppose* $q > k-2$. *For any* $\varepsilon > 0$ *there exists a number* $\gamma_\varepsilon > 0$ *such that*

$$T(\gamma) < (1+\varepsilon) \frac{k-1}{q-k+2} \cdot \frac{1}{k_1} \gamma^{5-2k+q}$$

if $\gamma > \gamma_\varepsilon$.

To improve on this estimate and also to obtain a lower bound, we use the identity

$$H(t) = \frac{1}{2k-2}\int_t^{\infty} s^{1-k}\{(2k-3)f(y(s))-y(s)f'(y(s))\}y'(s)ds, \qquad (15)$$

where

$$H(t) = \tfrac{1}{2}ty'^2 - \tfrac{1}{2}yy' + \frac{1}{2k-2}t^{1-k}yf(y).$$

It yields, if we set $t = T$:

$$Ty'^2(T) = \int_T^{\infty} s^{-k}g(y(s))ds,$$

where

$$g(y) = \frac{2k-3-q}{2(q+1)}\,y^{q+1}.$$

By estimating both $y'(T,\gamma)$ and the integral in (15) for large values of γ we obtain the following asymptotic estimate.

<u>THEOREM 3.</u> *Suppose* $q > 1$, $k > 2$ *and* $q > k-2$. *Then*

$$\lim_{\gamma\to\infty} \gamma^{2k-5-q}\,T(\gamma) = A(k,q),$$

where

$$A(k,q) = \frac{(k-1)^{\frac{k-3}{k-2}}}{k-2}\cdot\frac{2k-3-q}{q+1}\cdot\frac{\Gamma(\frac{q}{k-2}-1)\,\Gamma(\frac{1}{k-2}+1)}{\Gamma(\frac{q+1}{k-2})}$$

and Γ *denotes the gamma function.*

<u>EXAMPLE.</u> Let $N = 3(\Rightarrow k = 4)$ and $q = 3$. Then

$$\lim_{\gamma\to\infty} T(\gamma) = A(4,3) = \frac{1}{8}\pi\sqrt{3}. \qquad (16)$$

Thus problem III has a solution for every $T \in (0, \frac{1}{8}\pi\sqrt{3})$. By an entirely different method, this result was also found by Brezis [2].

In Fig. 2a we sketch the graph of $T(\gamma)$ when $N = 3$ and $q = 2.5$, 3 and 4. In Fig. 2b we do the same for the corresponding function $R(\gamma)$.

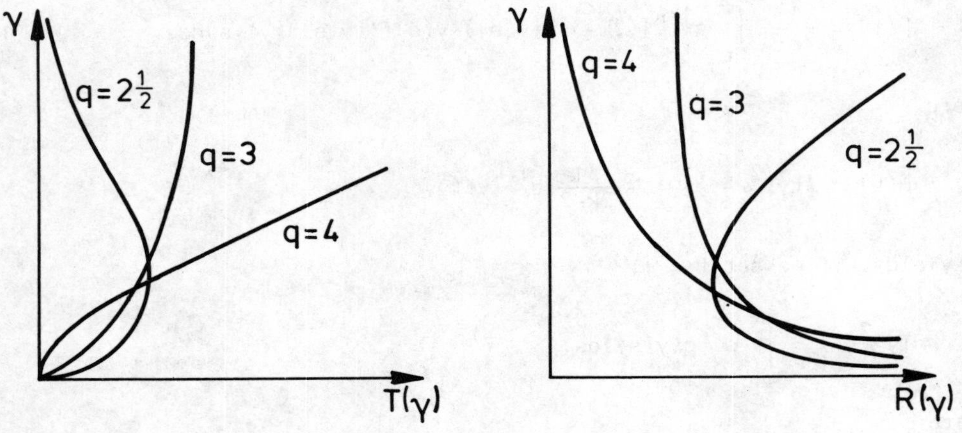

Fig. 2. The graphs of $T(\gamma)$ and $R(\gamma)$ for $N = 3$ and $q = 2.5, 3$ and 4.

It is clear from Theorem 3 that if

$$1 < q < 2k-5 = \frac{6-N}{N-2} \tag{17}$$

then

$$T(\gamma) \to 0 \quad \text{and} \quad R(\gamma) \to \infty \text{ when } \begin{matrix} \gamma \to 0 \\ \gamma \to \infty \end{matrix} .$$

Thus, in that case there exists a number R^* such that if $R^* < R < \infty$, Problem I has at least two solutions, partially confirming a conjecture of Brezis and Nirenberg [3].

Note that as in the case $q = 1$, $N = 4$ is a critical value. Indeed, if $N = 4$, then $k = 3$ and $2k - 5 = 1$. Thus, only if $2 < N < 4$ can (17) be satisfied.

REFERENCES

1. Atkinson, F.V. and L.A. Peletier: Emden-Fowler equations involving critical exponents, Nonlinear Anal. TMA (to appear).
2. Atkinson, F.V. and L.A. Peletier: On perturbation of critical growth rates in Emden-Fowler equations (to appear).
3. Brezis, H.: private communication.
4. Brezis, H. and L. Nirenberg: Positive solutions of nonlinear elliptic equations involving critical Sobolev exponents, Comm. Pure Appl. Math. 36 (1983), 437-477.

5. Fowler, R.H.: The form near infinity of real continuous solutions of a certain differential equation of the second order, Quart. J. Math. Cambridge Ser. 45 (1914), 289-350.
6. Fowler, R.H.: Further studies of Emden's and similar differential equations, Quart. J. Math. Oxford Ser. 2 (1931), 259-288.
7. Gidas, B., Wei-Ming Ni and L. Nirenberg: Symmetry and related properties via the maximum principle, Comm. Math. Phys. 68 (1979), 209-243.

F.V. Atkinson
Department of Mathematics
University of Toronto
Toronto, Ontario M5S 1A1
Canada

L.A. Peletier
Rijksuniversiteit te Leiden
Subfaculteit der Wiskunde en
Informatica
Wassenaarseweg 80
Postbus 9512
2300 RA Leiden
Nederland

C BANDLE

On a reaction-diffusion problem with a dead core

I) A simple model for an isothermal chemical reaction leads to the following nonlinear parabolic problem

$$\begin{cases} u_t(x,t) - \Delta u(x,t) = -\lambda f(u) & \text{in } D \times \mathbb{R}^+, \quad \lambda > 0 \\ u(x,t) = \gamma & \text{in } \partial D \times \mathbb{R}^+, \quad \gamma \in (0,1] \\ u(x,0) = u_0(x), & \text{in } D, \quad 0 < u_0(x) \leq 1 \end{cases} \tag{1}$$

Here $D \subset \mathbb{R}^N$ is the vessel where the reaction takes place, $u = u(x,t)$ stands for the concentration of the reactant and f is the production density satisfying the conditions

$$f \in C^p(\mathbb{R}^+ \cup \{0\}) \cap C^1(\mathbb{R}^+) \quad \text{for some } p \in (0,1); \tag{A-1}$$

$$f' \geq 0, \quad f(0) = 0. \tag{A-2}$$

Problems of this type have been investigated by a number of authors ([1]-[4], [6]-[10]).

Different approaches have been used to establish the following

<u>EXISTENCE THEOREM</u> ([1], [5], [10]). *If* $u_0 \in C^p(D)$, *then* (1) *has a unique classical solution.*

The uniqueness follows from (A-2). It should be mentioned that von Wahl's proof [10] applies also to functions f with a finite jump at the origin, such as $f(t) = \begin{cases} 0 & \text{for } t = 0 \\ 1 & \text{for } t > 0 \end{cases}$.

II) One feature of special interest is the so-called <u>dead core</u> $\Omega(t) := \{x \in D : u(x,t) = 0\}$, the subregion of D where no reaction takes place.

Let $u_\infty(x) = \lim_{t\to\infty} u(x,t)$ be the stationary state and let Ω_∞ be its corresponding dead core. If u_0 satisfies

$$-\Delta u_o \geq -\lambda f(u_o), \tag{B-1}$$

then u(x,.) is non-increasing and we therefore have

$$\Omega(t) \subseteq \Omega(t+\Delta t) \subseteq \ldots \subseteq \Omega_\infty, \quad \Delta t > 0. \tag{2}$$

The next monotonicity properties are obtained by means of the method of upper and lower solutions.

LEMMA 1. [1].
(i) *If* $\lambda_1 \geq \lambda_2$, *then* $\Omega_{\lambda_1}(t) \supseteq \Omega_{\lambda_2}(t)$.

(ii) *If* $D_1 \supseteq D_2$, *then* $\Omega_{D_1}(t) \supseteq \Omega_{D_2}(t)$.

It is clear that for small values of λ and t, $\Omega(t) = \emptyset$.

LEMMA 2. [3]. *There exists* $\lambda_o > 0$ *such that* $\Omega_\infty = \begin{cases} \emptyset & \text{for } \lambda < \lambda_o \\ \neq \emptyset & \text{for } \lambda \geq \lambda_o \end{cases}$. *If* $\Omega_\infty = \emptyset$ *for all* $\lambda \in \mathbb{R}^+$, *we set* $\lambda_o = \infty$.

Suppose that D lies in a strip of height 2d. Then, by comparing the two dead cores, we obtain [3],

$$\sqrt{\lambda_o} \geq (\sqrt{2}d)^{-1} \int_0^\gamma \frac{ds}{\sqrt{F(S)}}, \quad F(u) := \int_0^u f(s)\, ds. \tag{3}$$

A lower bound is constructed as follows.

From Lemma 1-(ii) we have $\lambda_o(D) \leq \lambda_o(B)$ for any sphere B contained in D, in particular for the insphere B_i of radius r_i. If v(x) solves

$$v''(x) = \lambda f(v(x)) \quad \text{in} \quad (-r_i, r_i), \quad v(-r_i) = v(r_i) = \gamma$$

then v(r), r > 0 satisfies, in view of (A-2)

$$v''(r) + \frac{N-1}{r} v'(r) \leq Nv''(r) = \lambda N f(v).$$

It is therefore an upper solution for the stationary case of (1) in B_i with λ replaced by $N\lambda$. From these considerations we get

$$\sqrt{\lambda_0} \le (\sqrt{2}\, r_i)^{-1} N \int_0^{\gamma} \frac{ds}{\sqrt{F(s)}} \tag{4}$$

and

<u>COROLLARY 1.</u> *If* f *is differentiable at zero, then* $\lambda_0 = \infty$.

Further estimates for λ_0 can be found in [3, 1, 6, 9]. Friedman and Phillips [8] also studied the convexity of Ω_∞ in the case of convex domains.

III) Consider problem (1) with initial data u_0 satisfying (B-1). In view of the monotonicity of $\Omega(t)$ (see (2)) it makes sense to define the onset time τ

$$\tau := \inf\{t : \Omega(t) \neq \emptyset\}.$$

<u>LEMMA 3.</u> *If* $u_0(x) \ge c_0$ $(x \in D)$, *then* $\tau \ge \int_0^{c_0} \{\lambda f(s)\}^{-1} ds =: \tau_0$.

<u>Proof.</u> $\underline{u}(x,t) = z(t)$ with $\dot{z} = -\lambda f(z)$, $z(0) = c_0$ is a lower solution for (1). The estimate follows from the fact that $z(\tau_0) = 0$.

Upper bounds for τ are obtained by means of suitable upper solutions. The following choice has been proposed in [1]:

$$\overline{u}(x,t) = z(\alpha t v(x)),$$

where v is any function, positive in some subdomain $D' \subseteq D$ with $v = 0$ on $\partial D'$, $\alpha \in \mathbb{R}^+$ and where z is such that $\dot{z} = -f(z)$, $z(0) = \max\{\sup_x u_0(x), \gamma\} =: c_1$.

A straightforward calculation yields

$$\overline{u}_t - \Delta\overline{u} = -f(\overline{u})\{\alpha v - \alpha t \Delta v + (\alpha t)^2 f'(\overline{u})|\nabla v|^2\} \quad \text{in } D' \times \mathbb{R}^+. \tag{5}$$

As before we have $z(\tau_1) = 0$ for $\tau_1 = \int_0^{c_1} \frac{ds}{f(s)}$. $\overline{u}$ is therefore only defined for $\alpha t v_{max} \le \tau_1$. In the sequel we shall make the additional assumption

$$f(t) = t^p f_0(t), \quad f_0(0) \neq 0, \quad f_0 \in C^2, \quad p \in [0,1). \tag{A-3}$$

Without loss of generality we may require that

$$f_o(t) \leq 1. \tag{6}$$

Under these assumptions any upper solution for the problem

$$\begin{cases} u_t - \Delta u = -\lambda u^p & \text{in} \quad D' \times \mathbb{R}^+ \\ \qquad u = c_1 & \text{in} \quad \partial D' \times \mathbb{R}^+ \\ u(x,0) = c_1 & \text{in} \quad D' \end{cases} \tag{7}$$

is an upper solution for (1) in $D' \times \mathbb{R}^+$.

By taking for D' the insphere B_i and for v the radially symmetric function $v = r_i^2 - r^2$ we get

<u>THEOREM 1.</u> *Let the conditions* (A-1) - (A-3) *hold and suppose that*

$$\lambda > \frac{1}{r_i^2}\left(\frac{c_1^{1-p}}{(1-p)^2}(2N(1-p) + 4p)\right) =: \frac{1}{r_i^2} k.$$

Then

$$\tau < \frac{c_1^{1-p}}{(1-p)[\lambda - k/r_i^2]} .$$

The same arguments yield

$$\text{dist } \{\Omega(t), \partial D\} \leq \sqrt{k\Big/\left(\lambda - \frac{c_1^{1-p}}{(1-p)t}\right)} . \tag{8}$$

In the case of convex domains we have the sharp estimate [3, 6, 8]

$$\frac{1}{\sqrt{2\lambda}} \int_0^{c_o} \frac{ds}{\sqrt{F(s)}} \leq \text{dist } \{\Omega_\infty, \partial D\}, \tag{9}$$

which according to (2) is also a lower bound for dist $\{\Omega(t), \partial D\}$. Friedman and Phillips [8] showed that for λ sufficiently large

$$\frac{\tilde{\gamma}}{\sqrt{\lambda}} - \frac{\tilde{c}_1}{\lambda} \leq \text{dist } \{\Omega_\infty, \partial D\} \leq \frac{\tilde{\gamma}}{\sqrt{\lambda}} + \frac{\tilde{c}_1}{\lambda} , \quad \tilde{c}_1, \tilde{\gamma} > 0.$$

Different results are obtained by varying the function v. For instance, if v is a solution of

$$\Delta v + q(v) = 0 \quad \text{in} \quad D, \quad v = 0 \quad \text{on} \quad \partial D, \quad q \geq 0 \tag{10}$$

and if we use gradient bounds of Payne and Stakgold [1] we find

THEOREM 2. *Let* ∂D *have positive mean curvature and* $v_m = v_{max}$, $q_m = \max_{0 \leq v \leq v_m} q(v)$. *Under the conditions* (A-1) - (A-3) *there exists a dead core for each* λ *satisfying* $\lambda \geq \alpha v_m + \dfrac{q_m c_1^{1-p}(1+p)}{v_m(1-p)^2}$ *for some* $\alpha > 0$, *with time of onset* $\tau \leq \dfrac{1}{\alpha v_m (1-p)}$.

IV) Suppose throughout this part that u_0 satisfies (B-1). The question arises, whether for any $x_0 \in \Omega_\infty(\lambda)$ there exists a time $\tau(x_0)$ such that $x_0 \in \Omega_\infty(t)$ for all $t \geq \tau(x_0)$.

THEOREM 3. [1]. *If* $x_0 \in \partial\Omega_\infty(\lambda)$, *then* $\tau(x_0) = \infty$.

Let now $x_0 \in \text{int}\ \Omega_\infty(\lambda)$ and define $\lambda^* := \inf \{\lambda : x_0 \in \Omega_\infty(\lambda)$. Since [8] $\text{dist}\ \{\partial\Omega_\infty(\lambda), \partial\Omega_\infty(\lambda')\} > 0$ for $\lambda' > \lambda$, we have $\lambda^* < \lambda$.

THEOREM 4. *Under the assumptions* (A-1) *and* (A-2) *we have*

$$\tau(x_0) < \frac{1}{\lambda-\lambda^*}\int_0^{c_1} \frac{dz}{f(z)} < \frac{1}{\lambda-\bar{\lambda}}\int_0^{c_1} \frac{dz}{f(z)}$$, *where* $\bar{\lambda}$ *is any upper bound for* λ^*.

EXAMPLE. $\bar{\lambda}^{1/2} = (\sqrt{2}\ \text{dist}\ \{x_0, \partial D\})^{-1} N \displaystyle\int_0^{c_1} \frac{dz}{F(z)}$ (see (4)).

Proof. Let u_∞ be the stationary solution of (1) with $\lambda = \lambda^*$ and $\dot{z} = -(\lambda-\lambda^*)f(z)$, $z(0) = c_1$. Then, setting $w = u_\infty + z$, we get

$$w_t - \Delta w = -(\lambda-\lambda^*)f(z) - \lambda^* f(u_\infty) \geq -\lambda f(w).$$

The assertion now follows.

Other estimates of this type are found in [1], [2].

V) Notice that the same techniques of constructing upper and lower solutions apply to the fast diffusion problem where Δu is replaced by $\Delta\varphi(u)$, φ being a monotone function. If we put $w = \varphi(u)$, $u = \psi(w)$ the problem takes the form

$$\begin{cases} \psi'(w)w_t - \Delta w = -\lambda f[\psi(w)] =: -\lambda g(w) & \text{in} \quad D \times \mathbb{R}^+ \\ w = \varphi(\gamma) & \text{in} \quad \partial D \times \mathbb{R}^+ \\ w(x,0) = \varphi[u_o(x)] & \text{in} \quad D. \end{cases} \tag{11}$$

By considering the upper solution $\overline{u} = z(\alpha t v(x))$ with $\psi'(z)\dot{z} = -g(z)$, $z(0) = \varphi(c_1)$, the same reasoning as for Theorem 1 or 2 yields

COROLLARY 2. *If* $\int_0^{\varphi(c_1)} \frac{\psi'(z)}{g(z)}\, dz < \infty$, *then for sufficiently large* λ, *there exists a dead core which starts to form after finite time.*

All the other results in sections II) - IV) extend after suitable modifications to problem (11).

REMARK. Further applications of this technique are problems of quenching and problems whose solutions blow up.

REFERENCES

1. Bandle, C. and I. Stakgold: The formation of the dead core in parabolic reaction-diffusion equations, Trans. Amer. Math. Soc. 286 (1984), 275-293.
2. Bandle, C. and I. Stakgold: Reaction-diffusion and dead cores. In "Free boundary problems: applications and theory", Vol. III and IV (Pitman, 1985).
3. Bandle, C., R. Sperb and I. Stakgold: Diffusion and reaction with monotone kinetics, Nonlinear Anal. TMA 8 (1984), 321-333.
4. Bandle, C.: A note on optimal domains in reaction-diffusion problems, Z. Angew. Anal. (to appear).
5. Deuel, J. and P. Hess: Nonlinear parabolic boundary value problems with upper and lower solutions, Israel J. Math. 29 (1978), 92-104.
6. Diaz, J.I. and J. Hernandez: On the existence of a free boundary for a class of reaction-diffusion systems, SIAM J. Math. Anal. 15 (1984), 670-685.
7. Diaz, J.I. and J. Hernandez: Some results in the existence of free boundaries for parabolic reaction-diffusion systems. In "Trends in Theory and Practice of Nonlinear Differential equations", V. Lakshmikantham Ed., pp. 149-156 (Dekker, 1984).
8. Friedman, A. and D. Phillips: The free boundary of a semilinear elliptic equation, Trans. Amer. Math. Soc. 282 (1984), 153-182.

9. Stakgold, I.: Estimates for some free boundary problems. In "Ordinary and Partial Differential Equations", Lecture Notes in Mathematics, vol. 846 (Springer, 1981).

10. von Wahl, W.: On parabolic systems with discontinuous non-linearities (to appear).

C. Bandle
Mathematisches Institut
Universität Basel
Rheinsprung 21
CH-4051 Basel
Switzerland

A BENSOUSSAN

On some parabolic equations arising from ergodic control

Introduction

The objective of this article is first to study equations of the type

$$-\frac{\partial\varphi}{\partial\tau} - \Delta_y\varphi + \rho = F(y,\tau,D\varphi) \tag{1}$$

where

$$F(y,\tau,q) = \operatorname*{Inf}_v \{\psi(y,\tau,v) + q.g(y,\tau,v)\}$$

is a Hamiltonian. The solution is periodic in y and τ (all data have the same periodicity, period 1 in each component of y and τ_0 in τ). The constant ρ in (1) is an unknown as is φ. Note also that φ is defined up to an additive constant.

Equation (1) arises in solving a problem of stochastic ergodic control. The constant ρ has the following interpretation. Let us consider the stochastic differential equation

$$dy = g(y(\sigma),\sigma,v(\sigma))d\sigma + \sqrt{2}\, db(\sigma)$$
$$y(\tau) = y$$

where v(.) is a control. Then

$$\rho = \operatorname*{Min}_{v(.)} \varliminf_{T\to\infty} \frac{1}{T} E \int_\tau^T \psi(y(\sigma),\sigma,v(\sigma))d\sigma. \tag{2}$$

Besides their intrinsic interest, equations of the type (1) are connected to the solution of singular perturbation problems of the type

$$-\frac{\partial\varphi^\varepsilon}{\partial t} - \Delta_x\varphi^\varepsilon - \frac{1}{\varepsilon}\Delta_y\varphi^\varepsilon = F(x,y,t,\frac{t}{\varepsilon}, D_x\varphi^\varepsilon, \frac{1}{\varepsilon}D_y\varphi^\varepsilon) \tag{3}$$

$\varphi^\varepsilon = 0$ for $x \in \Gamma$, $\forall y,t$

φ^ε periodic in y

$\varphi^\varepsilon(x,y,T) = 0$

where $x \in \mathcal{O}$, a smooth bounded domain of R^n, whose boundary is denoted Γ.

The function F is a Hamiltonian of the form

$$F(x,y,t,\tau,p,q) = \operatorname*{Inf}_{v} [\ell(x,y,t,\tau,v) + p.f(x,y,t,\tau,v) + q.g(x,y,t,\tau,v)]. \tag{4}$$

Writing formally

$$\varphi^{\varepsilon}(x,y,t) = \varphi(x,t) + \varepsilon\varphi_1(x,y,t,\frac{t}{\varepsilon})$$

one obtains by equating the terms of order 0 in both sides of (3), the relation

$$-\frac{\partial\varphi_1}{\partial\tau} - \Delta_y\varphi_1 - \frac{\partial\varphi}{\partial t} - \Delta_x\varphi = F(x,y,t,\tau,D_x\varphi,D_y\varphi_1). \tag{5}$$

Equation (5) must be considered as an equation in φ_1, which is a function of y,τ, and φ,x,t are parameters. Hence (5) is of the type (1).

Setting

$$\psi(y,\tau,v) = \ell(x,y,t,\tau,v) + p.f(x,y,t,\tau,v)$$

where x,t,p are parameters, the study of (1) introduces a constant ρ (with respect to y,τ but depending on x,t,p) such that

$$-\frac{\partial\varphi_1}{\partial\tau} - \Delta_y\varphi_1 + \rho(x,t,p) = \operatorname*{Inf}_{v} \{\ell(x,y,t,\tau,v) + p.f(x,y,t,\tau,v) + D_y\varphi_1.g(x,y,t,\tau,v)\}.$$

Identifying with (5) yields that φ must be the solution of the equation

$$-\frac{\partial\varphi}{\partial t} - \Delta_x\varphi = \rho(x,t,D\varphi). \tag{6}$$

Justifying the convergence of φ^{ε} to φ is the second objective of this article.

These problems generalize to the evolution case problems studied in A. Bensoussan [1] for the elliptic case.

1. Setting of the problem

1.1 Assumptions. Notation

Let

$$g(y,\tau,v) : R^d \times [0,\tau_o] \times U \to R^d \tag{1.1}$$
$$\psi(y,\tau,v) : R^d \times [0,\tau_o] \times U \to R$$

be functions periodic in y (period 1 in each component) and τ (period τ_o);
U metric space

$$U_{ad} \text{ compact subspace of } U \tag{1.2}$$

$$g,\psi \text{ continuous functions.} \tag{1.3}$$

Let us set

$$H(y,\tau,v,q) = \psi(y,\tau,v) + q.g(y,\tau,v) \tag{1.4}$$

$$F(y,\tau,q) = \operatorname*{Inf}_{v\in U_{ad}} H(y,\tau,v,q). \tag{1.5}$$

One calls H the Hamiltonian. The assumptions imply the existence of an optimal feedback $\hat{V}(y,\tau,q)$ such that

$$F(y,\tau,q) = H(y,\tau,\hat{V},q), \tag{1.6}$$

and $\hat{V}$ is measurable with values in U_{ad}.

1.2 The problem

To find $\varphi(y,\tau)$ and a constant ρ such that

$$\varphi \text{ periodic in } y,\tau \tag{1.7}$$

$$\varphi \in W^{2,1,p}(Y \times (0,\tau_o))$$

$$-\frac{\partial\varphi}{\partial\tau} - \Delta_y\varphi + \rho = F(y,\tau,D_y\varphi). \tag{1.8}$$

We have used the notation $Y = (0,1)^d$, and

$$W^{2,1,p}(Y \times (0,\tau_o)) = \{z \in L^p(Y \times (0,\tau_o)), \frac{\partial z}{\partial \tau}, \frac{\partial z}{\partial y_i}, \frac{\partial^2 z}{\partial y_i \partial y_j} \in L^p\}.$$

Then the first result is the following

THEOREM 1.1. *Under the assumptions* (1.1), (1.2), (1.3) *there exists a solution* φ, ρ of (1.7), (1.8). *This solution is unique up to an additive constant for* φ.

2. Main ideas of the proof of Theorem 1.1

2.1 Approximation

One considers the problem

$$-\frac{\partial \varphi_\alpha}{\partial \tau} - \Delta_y \varphi_\alpha + \alpha \varphi_\alpha = F(y,\tau,D_y\varphi_\alpha) \tag{2.1}$$

$$\varphi_\alpha \in W^{2,1,p}(Y \times (0,\tau_o))$$

φ_α periodic in y,τ.

For $\alpha > 0$, one first proves the existence and uniqueness of the solution of (2.1).

2.2 Fredholm alternative for a parabolic operator

Consider any feedback $v(y,\tau)$ (any measurable function from $Y \times (0,\tau_o)$ in U_{ad}). Set $g^v = g(y,\tau,v(y,\tau))$. One searches a function p^v solution of

$$\frac{\partial p}{\partial \tau} - \Delta_y p + \operatorname{div}(pg^v) = 0 \tag{2.2}$$

$$p \in L^2(0,\tau_o:H^1(Y))$$

$$\frac{\partial p}{\partial \tau} \in L^2(0,\tau_o;H^{-1}(Y))$$

p periodic in y,τ.

As a consequence of the Fredholm alternative in an adequate setting one proves the existence of a solution of (2.2). This solution is unique up to a multiplicative constant.

In fact additional important properties can be proven. Firstly

$$\int_Y p(y,\tau)dy = 1, \quad \forall\tau \tag{2.3}$$

and this sets the function. Moreover there exists positive constants δ_0, δ_1, such that

$$0 < \delta_0 \leq p^v(y,\tau) \leq \delta_1 \tag{2.4}$$

and the constants δ_0, δ_1 are independent of the feedback v(.).

2.3 Convergence

Setting

$$v_\alpha(y,\tau) = \hat{V}(y,\tau,D_y\varphi_\alpha)$$

one can rewrite (2.1) as

$$-\frac{\partial\varphi_\alpha}{\partial\tau} - \Delta_y\varphi_\alpha + \alpha\varphi_\alpha = \psi(y,\tau,v_\alpha) + D_y\varphi_\alpha.g(y,\tau,v_\alpha). \tag{2.5}$$

Consider p_α solution of (2.2) corresponding to $v = v_\alpha$, then from (2.4) one has

$$\delta_0 \leq p_\alpha(y,\tau) < \delta_1. \tag{2.6}$$

Set next

$$\rho_\alpha = \alpha\bar{\varphi}_\alpha = \frac{\alpha}{\tau_0}\int_Y\int_0^{\tau_0}\varphi_\alpha(y,\tau)dyd\tau$$

and

$$\tilde{\varphi}_\alpha = \varphi_\alpha - \bar{\varphi}_\alpha.$$

One first proves that $\tilde{\varphi}_\alpha$ remains bounded in $H^1(Y \times (0,\tau_0))$. This is obtained by successively testing (2.5) with $P_\alpha\tilde{\varphi}_\alpha$ and $\partial\tilde{\varphi}_\alpha/\partial\tau$, making use of the properties of p_α.

Note that ρ_α is bounded. In fact $\tilde{\varphi}_\alpha$ remains bounded in $W^{2,1,p}$. By extracting a subsequence one obtains a solution φ,ρ of (1.8).

The uniqueness relies on the interpretation of ρ, explicitly given by the formula

$$\rho = \underset{v(.)}{\text{Inf}} \frac{1}{\tau_o} \int_0^{\tau_o} \psi(y,\tau,v(y,\tau))p^v(y,\tau)dyd\tau. \tag{2.7}$$

3. Singular perturbations

3.1 Assumptions. Notations

Consider the functions

$$\begin{aligned} f(x,y,t,\tau,v) &: R^n \times R^d \times [0,T] \times [0,\tau_o] \times U \to R^n \\ g(x,y,t,\tau,v) &: R^n \times R^d \times [0,T] \times [0,\tau_o] \times U \to R^d \\ \ell(x,y,t,\tau,v) &: R^n \times R^d \times [0,T] \times [0,\tau_o] \times U \to R \end{aligned} \tag{3.1}$$

periodic in y, with period 1 in all components, periodic in τ with period τ_o, continuous

$$U_{ad} \text{ compact subset of } U. \tag{3.2}$$

Let us set

$$H(x,y,t,\tau,v,p,q) = \ell(x,y,t,\tau,v) + p.f(x,y,t,\tau,v) + q.g(x,y,t,\tau,v) \tag{3.3}$$

and

$$F(x,y,t,\tau,p,q) = \underset{v\in U_{ad}}{\text{Inf}} H(x,y,t,\tau,v,p,q). \tag{3.4}$$

Let $\mathcal{O}$ be a smooth bounded domain of R^n, whose boundary is denoted Γ. Since we shall consider only the values of x in $\mathcal{O}$, one can always assume that f,g,ℓ are bounded.

Let us consider the non linear parabolic problem

$$- \frac{\partial \varphi_\varepsilon}{\partial t} - \Delta_x \varphi^\varepsilon - \frac{1}{\varepsilon} \Delta_y \varphi^\varepsilon = F(x,y,t, \frac{t}{\varepsilon}, D_x \varphi^\varepsilon, \frac{1}{\varepsilon} D_y \varphi^\varepsilon) \tag{3.5}$$

$\varphi^\varepsilon = 0$ for $x \in \Gamma$, $\forall\, y,t$

φ^ε periodic in y

$\varphi^\varepsilon(x,t,T) = 0$

$\varphi^\varepsilon \in W^{2,1,p}(\mathcal{O} \times Y \times (0,T))$.

Let also $\hat{V}(x,y,t,\tau,p,q)$ be a Borel function with values in U_{ad}, which satisfies the infimum in (3.4). We shall set

$$u^\varepsilon(x,y,t,\tau) = \hat{V}(x,y,t, \frac{t}{\varepsilon}, D_x \varphi^\varepsilon, \frac{1}{\varepsilon} D_y \varphi^\varepsilon) \tag{3.6}$$

which is an optimal feedback in (3.5).

3.2 Description of the limit problem

Let $v(y,\tau)$ be a feedback (possibly depending on parameters $x,t,p,\ldots$). Let us denote

$$g^v(y,\tau) = g(x,t,y,\tau;v(y,\tau)). \tag{3.7}$$

Consider then the function $m^v(x,t;y,\tau)$ (in which x,t are parameters), solution of the problem

$$\frac{\partial m}{\partial \tau} - \Delta_y m + \operatorname{div}(m g^v) = 0 \tag{3.8}$$

$m \in L^2(0,\tau_o;H^1(Y))$

$\frac{\partial m}{\partial \tau} \in L^2(0,\tau_o;H^{-1}(Y))$

m periodic in y,τ

$\int_Y m\,dy = 1, \quad \forall\, x,t;\tau.$

There exists a unique solution of (3.8) satisfying

$$0 < \delta_0 \leq m^v(x,t;y,\tau) \leq \delta_1. \tag{3.9}$$

Let us set next

$$\rho(x,t,p) = \operatorname*{Inf}_{v(.)} \frac{1}{\tau_o} \int_0^{\tau_o} \int_Y m^v(x,t;y,\tau) \tag{3.10}$$

$$[\ell(x,y,t,\tau,v(y,\tau))+p.f(x,y,t,\tau,v(y,\tau))]dyd\tau.$$

The function ρ is u.s.c. and Lipschitz in p. Therefore one can consider the problem

$$-\frac{\partial\varphi}{\partial t} - \Delta\varphi = \rho(x,t;D\varphi) \tag{3.11}$$

$$\varphi|_\Gamma = 0, \quad \varphi(x,T) = 0$$

$$\varphi \in W^{2,1,p}(\mathcal{O} \times (0,T)), \quad \forall\, p \text{ with } 2 \leq p < \infty.$$

The main result is the following

THEOREM 3.1. *Under the assumptions* (3.1), (3.2), *one has*

$$\varphi^\varepsilon \to \varphi \text{ in } L^2(0,T;H^1(\mathcal{O} \times Y)) \text{ and} \tag{3.12}$$

$$\text{in } L^\infty(0,T;L^2(\mathcal{O} \times Y))$$

$$\text{if } \varepsilon = \frac{T}{N\tau_0}, \quad N \to \infty.$$

The condition on ε is useful for technical reasons, but is probably not essential.

Note that (3.11) is a Bellman equation, since ρ can be written

$$\rho(x,t,p) = \operatorname*{Inf}_{v(.)} [\tilde{\ell}(x,t;v(.))+p.\tilde{f}(x,t;v(.))].$$

3.3 Sketch of the proof

Let $v(y,\tau)$ be a feedback. Consider the solution $m_\varepsilon = m_\varepsilon^v$ of the problem:

$$\varepsilon \frac{\partial m_\varepsilon}{\partial t} - \varepsilon\Delta_x m_\varepsilon - \Delta_y m_\varepsilon + \operatorname{div}_y(m_\varepsilon g_\varepsilon^v) = 0 \tag{3.13}$$

$$\frac{\partial m_\varepsilon}{\partial \nu}\Big|_\Gamma = 0$$

m_ε periodic in y,t (in t the period is T)

$m_\varepsilon \in L^2(0,T;H^1(\mathcal{O} \times Y))$

where g_ε^v denotes the function

$$g_\varepsilon^v(x,t,y) = g(x,t,y, \frac{t}{\varepsilon}, v(y, \frac{t}{\varepsilon})). \tag{3.14}$$

There exists one and only one solution of (3.13) such that

$$\frac{1}{T} \frac{1}{|\mathcal{O}|} \int_0^T \int_{\mathcal{O}} \int_Y m_\varepsilon^v(x,t,y) dxdtdy = 1. \tag{3.15}$$

In fact one has also the properties

$$\int_Y m_\varepsilon^v(x,t,y)dy = 1, \quad \forall\, x,t \tag{3.16}$$

$$0 < \delta_0 \leq m_\varepsilon^v(x,t,y) \leq \delta_1 \quad \text{a.e.}$$

Considering $m_\varepsilon = m_\varepsilon^{u^\varepsilon}$ (i.e. the feedback $v = u^\varepsilon$ defined by (3.6)), and multiplying (3.5) by $m_\varepsilon \varphi^\varepsilon$ yields some energy estimates, namely

$$\int_0^T \int_{\mathcal{O}} \int_Y |D_x\varphi^\varepsilon|^2 dxdydt \leq C, \tag{3.17}$$

$$\int_0^T \int_{\mathcal{O}} \int_Y |D_y\varphi^\varepsilon|^2 dxdydt \leq C\varepsilon$$

$$|\varphi^\varepsilon| \leq \frac{\|\ell\|}{T}.$$

The next step consists in proving the following equality, valid for any $\varphi(x,t) \in L^2(0,T;H_0^1(\mathcal{O}) \cap H^2(\mathcal{O}))$ with $\frac{\partial\varphi}{\partial t} \in L^2(0,T;L^2(\mathcal{O}))$

$$\begin{aligned}
&\frac{1}{2} \iint_{\mathcal{O}\times Y} m_\varepsilon(\varphi^\varepsilon(x,y,t)-\varphi(x,t))^2 dxdy + \\
&\int_t^T \iint m_\varepsilon |D_x\varphi^\varepsilon - D_x\varphi|^2 dxdyds + \frac{1}{\varepsilon}\int_t^T \iint m_\varepsilon |D_y\varphi^\varepsilon|^2 dxdyds = \\
&\int_t^T \iint (\ell^\varepsilon + D_x\varphi^\varepsilon . f^\varepsilon) m_\varepsilon(\varphi^\varepsilon - \varphi) dxdyds + \frac{1}{2}\int \varphi^2(x,t)dx + \\
&\int_t^T \int |D_x\varphi|^2 dxds + \int_t^T \iint m_\varepsilon \varphi^\varepsilon \left(\frac{\partial\varphi}{\partial t} + \Delta\varphi\right) dxdyds.
\end{aligned} \tag{3.18}$$

It is then necessary to estimate $\partial\varphi^\varepsilon/\partial t$ and this is a delicate step. One proves that

$$\frac{\partial\varphi^\varepsilon}{\partial t} \text{ is bounded in } L^2(0,T;V'), \text{ where} \tag{3.19}$$

V = subspace of $H^1(\mathcal{O} \times Y)$ of functions which

vanish on Γ and are periodic in y.

This is achieved by writing

$$\varphi^\varepsilon(x,y,t) = \overline{\varphi}^\varepsilon(x,t) + \varepsilon\tilde{\varphi}^\varepsilon(x,y,t)$$

where

$$\overline{\varphi}^\varepsilon(x,t) = \int_Y \varphi^\varepsilon(x,y,t)m_\varepsilon^{u^\varepsilon}(x,y,t)dy$$

and proving separately that $\partial\overline{\varphi}^\varepsilon/\partial t \in L^2(0,T;H^{-1}(\mathcal{O}))$, and $\varepsilon\ \partial\tilde{\varphi}^\varepsilon/\partial t$ is bounded in $L^2(0,T;V')$.

One can then extract a subsequence such that

$$\varphi^\varepsilon \to \varphi \text{ in } L^2(0,T;H^1(\mathcal{O} \times Y)) \cap L^\infty \text{ weakly} \tag{3.20}$$

$$\frac{\partial\varphi^\varepsilon}{\partial t} \to \frac{\partial\varphi}{\partial t} \text{ in } L^2(0,T;V') \text{ weakly.}$$

Then φ is a function of x,t only, $\varphi \in L^2(0,T;H_0^1(\mathcal{O}))$, $\frac{\partial\varphi}{\partial t} \in L^2(0,T;H^{-1}(\mathcal{O}))$, $\varphi(x,T) = 0$.

Moreover one has

$$\varphi^\varepsilon \to \varphi \text{ in } L^2(0,T;H^1(\mathcal{O} \times Y)) \text{ strongly and} \tag{3.21}$$

$$L^\infty(0,T;L^2(\mathcal{O} \times Y)) \text{ strongly.}$$

To identify the limit pick any feedback $v(y,\tau)$, and derive from (3.5)

$$-\frac{\partial\varphi^\varepsilon}{\partial t} - \Delta_x\varphi^\varepsilon - \frac{1}{\varepsilon}\Delta_y\varphi^\varepsilon \le \ell(x,t,y,\tfrac{t}{\varepsilon},v(y,\tfrac{t}{\varepsilon}))$$
$$+ D_x\varphi^\varepsilon.f(x,t,y,\tfrac{t}{\varepsilon},v(y,\tfrac{t}{\varepsilon})) + \frac{1}{\varepsilon}D_y\varphi^\varepsilon.g(x,t,y,\tfrac{t}{\varepsilon},v(y,\tfrac{t}{\varepsilon})). \tag{3.22}$$

Let $\varphi(x) \in C_0^\infty(\mathcal{O})$, $\varphi \geq 0$, $\beta(t) \in C^1([0,T])$, $\beta(0) = 0$, $\beta(t) \geq 0$. Let m_ε^v be the solution of (3.13).

After some manipulations, one derives easily the inequality

$$\int_0^T \iint \varphi^\varepsilon m_\varepsilon^v \varphi\beta' dxdydt + 2 \int_0^T \iint m_\varepsilon^v D_x\varphi^\varepsilon . D_x\varphi\beta dxdydt \tag{3.23}$$

$$+ \int_0^T \iint m_\varepsilon^v \varphi^\varepsilon \Delta\varphi\beta dxdydt \leq \int_0^T \iint (\ell_\varepsilon^v + D_x\varphi . f_\varepsilon^v) m_\varepsilon^v \varphi\beta$$

$$+ \int_0^T \iint (D_x\varphi^\varepsilon - D_x\varphi) f_\varepsilon^v m_\varepsilon^v \varphi\beta dxdydt.$$

At this stage the hypothesis $\varepsilon = T/N\tau_0$, $N \to \infty$, becomes useful. It implies indeed that

$$m_\varepsilon^v(x,y,t) = \mu_\varepsilon^v(x,y,t;\ \frac{t}{\varepsilon})$$

where $\mu_\varepsilon^v(x,y,t,\tau)$ is the solution of

$$\varepsilon \frac{\partial\mu_\varepsilon}{\partial t} + \frac{\partial\mu_\varepsilon}{\partial\tau} - \varepsilon\Delta_x\mu_\varepsilon - \Delta_y\mu_\varepsilon + \mathrm{div}(\mu_\varepsilon g^v) = 0 \tag{3.24}$$

$$\frac{\partial\mu_\varepsilon}{\partial\nu}\Big|_\Gamma = 0$$

μ_ε periodic in y,τ,t (period τ_0 in τ and T in t).

The important step is that

$$\mu_\varepsilon^v \to m^v \quad \text{in} \quad L^2(\mathcal{O} \times Y \times (0,T) \times (0,\tau_0)) \tag{3.25}$$

where m^v is the solution of (3.8).

From this it is possible to deduce that

$$\int_0^T \iint (\ell_\varepsilon^v + D_x\varphi . f_\varepsilon^v) m_\varepsilon^v \varphi\beta dxdydt \tag{3.26}$$

$$\to \int_0^T dt \int_{\mathcal{O}} dx\varphi(x)\beta(t) \frac{1}{\tau_0} \int_Y dy \int_0^{\tau_0} d\tau [\ell^v + D_x\varphi . f^v] m^v$$

and thus it becomes possible to pass to the limit in (3.23), whence

$$- \frac{\partial\varphi}{\partial t} - \Delta\varphi \leq \rho(x,t,D\varphi).$$

Some relatively similar (but slightly more technical) steps allow us to obtain the reverse inequality, so that the desired result follows.

REFERENCES

1. Bensoussan, A.: Méthodes de perturbations en contrôle optimal (Dunod, to be published).

2. Bensoussan, A., J.L. Lions and G. Papanicolaou: Asymptotic Analysis for Periodic Structures (North-Holland, 1982).

A. Bensoussan
INRIA
Domaine de Voluceau-Rocquencourt
B.P. 105
78153 Le Chesnay CEDEX
France

M BIROLI

A Green function method to obtain an estimate on the modulus of continuity in a parabolic problem

1. Introduction and results

The goal of this paper is to describe a method to obtain an estimate of the modulus of continuity of a weak solution to parabolic equations or variational inequalities.

For the sake of simplicity we use here such a method to obtain an estimate of the modulus of continuity of a weak solution to a linear parabolic equation in terms of the so-called "Wiener integral".

We remark that the first results on this problem have been given in the elliptic case by Maz'ja ([11],[12]), dealing with the linear or with some quasilinear cases, or by Gariepy and Ziemer [8], in the general elliptic case.

In the parabolic case the Wiener criterion for the heat equation has been proved by Evans and Gariepy [6]; previous results in this direction have been obtained by Lanconelli [10].

In the nonlinear parabolic case a sufficient condition for the continuity at a boundary point has been given by Ziemer [15], Gariepy and Ziemer [9].

We observe that the results in [9], [15] are of qualitative type and no estimates on the modulus of continuity are obtained.

In recent studies concerning the continuity of the solution of an elliptic obstacle problem ([7],[13],[14]), Frehse and Mosco have introduced a method, which uses the Green function of the principal part of the involved elliptic operator, to obtain an estimate of the modulus of continuity in terms of the so-called "Wiener integral" relative to the obstacle.

This method, in the case of boundary points and of weak solutions of an elliptic equation, allows us to prove Maz'ja's result in [11] and also the result by Gariepy and Ziemer in [8], if the principal part of the operator is linear.

In the following we will describe an extension of this method to the parabolic case, obtaining an estimate of the modulus of continuity at a boundary point for a weak solution of a parabolic equation.

Some possible generalizations or applications are given at the end of this section.

Finally, we remark that the results considered here have been obtained jointly with Mosco ([2],[3]).

Let Q be an open bounded set in R^{N+1}, $N \geq 3$, $\Sigma = \partial Q$ and a_{ij} $(i,j=1,2,\ldots,N)$ be such that $a_{ij} \in L^\infty(Q)$ and

$$\Sigma_{i,j=1}^{N} a_{ij}(x,t)\,\xi_i\,\xi_j \geq \lambda|\xi|^2,\quad |a_{ij}(x,t)| \leq \Lambda \tag{1.1}$$

a.e. in Q, for any $\xi \in R^N$, where $\lambda > 0$.

A function $u \in V^2(Q) = L^\infty(R;L^2(\Omega_t)) \cap H^{0,1}(Q)$ $(\Omega_t = Q \cap \{\tau = t\})$ is a weak solution of our problem if

$$\int_Q \{(D_t\varphi)u + \Sigma_{i,j=1}^{N} a_{ij}\, D_{x_i} u\, D_{x_j}\varphi\}\, dxdt = 0 \tag{1.2}$$

for any $\varphi \in C_o^\infty(Q)$. We observe that from (1.1) we have easily $D_t u \in Z(Q) = \{v \in D'(Q) \,|\, v = \operatorname{div} g,\ g \in (L^2(Q))^N\}$; then

$$\langle D_t u,\varphi\rangle + \Sigma_{i,j=1}^{N} \int_Q a_{ij}\, D_{x_i} u\, D_{x_j}\varphi\ \ dxdt = 0 \tag{1.3}$$

for any $\varphi \in H_o^{0,1}(Q) = \{\text{closure of } C_o^\infty(Q) \text{ in } H^{0,1}(Q)\}$, where $\langle\, ,\rangle$ denotes the duality between $Z(Q)$ and $H_o^{0,1}(Q)$, and we have $u \in W^1(Q)=\{v \in H^{0,1}(Q) \,|\, D_t v \in Z(Q)\}$.

Let $z_o = (x_o,t_o) \in \Sigma$. We define:

(i) $u(z_o) \leq 1$ (weakly), $u \in W^1(Q)$, if for any $k > 1$ there exists $r > 0$ and a sequence $u_n \in W^{1,\infty}(Q)$ such that $u_n \to u$ in $W^1(Q)$ and $\operatorname{supp}\{\eta(u_n-k)^+\} \subset Q \cap Q(z_o,r)$ whenever $\eta \in C_o^\infty(Q(z_o,r))$ $(Q(z_o,r) = B(x_o,r)\times(t_o-r^2,t_o+r^2)$, $B(x_o,r) = \{|x-x_o| < r\})$.

The definition of:

(ii) $u(z_o) \geq 1$ (weakly) is analogous and $u(z_o) = 1$ (weakly) if both (i) and (ii) hold.

Now we give the notion of capacity (the Γ-capacity) which will be used in the following. Let E be a compact set in $Q(z_o,r)$; we define:

$$\Gamma\text{-cap}(E,Q(z_o,r)) = \inf\{\|v\|^2_{L^\infty(t_o-r^2,t_o+r^2;L^2(\Omega_t))} + \int_{Q(z_o,r)} |D_x v|^2 dxdt\},$$

where the infimum is taken over all functions in $V^2(Q)$ with $\operatorname{supp}(v) \subset Q(z_o,r)$ and $E \subset \operatorname{int}\{z \,|\, v(z) \geq 1\}$.

In the following we denote

$$\Delta_\vartheta(r) = \Gamma\text{-cap}(Q^c \cap Q^-_\vartheta(z_o,r), Q(z_o,2r)),$$

where

$$Q^-_\vartheta(z_o,r) = B(x_o,(1-\vartheta)^{\frac{1}{2}}r) \times (t_o-(1-\vartheta)r^2, t_o-\vartheta r^2)(\vartheta \in (0,1/16)),$$

and, if $\sigma_N = \Gamma\text{-cap}(Q(z_o,r), Q(z_o,2r))$,

$$\delta_\vartheta(r) = (\sigma_N r^N)^{-1} \Delta_\vartheta(r).$$

Finally we set

$$\omega_\vartheta(r) = \exp(-\int_r^1 \delta_\vartheta(\rho)\rho^{-1} d\rho).$$

<u>THEOREM 1.</u> *Let* f,g *be continuous functions,* $f(z_o) = g(z_o)$ *and* $g \le u \le f$ *q.e. in* $Q(z_o,\overline{R}) \cap \Sigma$.

We have

$$\text{osc}_{Q(z_o,r)} u \le C\omega_\vartheta(r)^{\beta(1-\alpha)} + \Psi_f(\omega_\vartheta(r)^\alpha) + \Psi_g(\omega_\vartheta(r)^\alpha).$$

for any $\alpha,\beta \in (0,1)$, $\omega_\vartheta(r)^\alpha \le \vartheta\overline{R}$, $\vartheta \in (0,\vartheta_o)$ ϑ_o *suitable depending on* λ, Λ, N, $C = C(\lambda, \Lambda, N, \vartheta_o)$, *where* Ψ_f *and* Ψ_g *are the moduli of continuity of* f,g.

<u>COROLLARY 1.</u> *Let* f,g *be as in Theorem 1 and Hölder continuous; if* $\omega_\vartheta(\rho) \le \rho^\nu$, $\nu \in (0,1)$, *we have*

$$\text{osc}_{Q(z_o,r)} u \le Cr^\gamma$$

where C *is as in Theorem* 1. *and* $\gamma \in (0,1)$ *depends on* α,β,ν *and on the Hölder exponents of* f,g.

We will give the sketch of the proof of Theorem 1 in Section 2. We observe that the result can be extended to the <u>nonlinear case with quadratic growth in the gradient</u>.

We now recall some other problems in which our method works and gives estimates on the modulus of continuity.

(1) The case of the modulus of continuity at an interior point of local

solutions of a parabolic obstacle problem (Biroli and Mosco [4]).

(2) The case of the modulus of continuity at boundary points of Q, which is supposed to be a cylinder, for solution of a two phases Stefan problem or of the porous medium equation (Biroli, Di Benedetto and Mosco [5]).

In those cases an estimate of the modulus of continuity by the so-called "Wiener integral" is also given, but a different notion of capacity is considered. Precisely, the capacity of a compact set E in $Q(z_0,r)$ is defined by

$$\mathrm{Cap}\,(E,Q(z_0,r)) = \inf\{\int_{Q(z_0,r)} |D_x v|^2 dxdt\},$$

where the infimum is taken over all functions v in the space $L^2(t_0-r^2,\ t_0+r^2;\ H_0^1(B(x_0,r))$ with $E \subset \mathrm{int}\{z|v(z) \geq 1\}$.

We observe that this notion of capacity is weaker than the capacity defined above.

2. Sketch of the proof of Theorem 1

The proof of Theorem 1 is based on a Poincaré's inequality involving only the spatial gradient, which is given for subsolutions of our problem. Precisely, we can prove the following result (here $\overline{z} = (\overline{x},\overline{t})$; $\|\cdot\|_r = \|\cdot\|_{L^2(B(\overline{x},r))}$).

LEMMA 1. (Poincaré's inequality) *Let* v *be a subsolution of our problem in* $Q(\overline{z},2r)$, *such that the following relation holds:*

$$\left| \|(v(t_2)-c)^+ \varphi\|_r^2 - \|(v(t_1)-c)^+ \varphi\|_r^2 \right| \leq \tag{2.1}$$

$$\leq K\{\int_{t_1}^{t_2}\int_{B(\overline{x},r)} [|D_x(v-c)^+|^2 \varphi^2 + |D_x(v-c)^+||D_x\varphi|\varphi(v-c)^+]dxdt\}$$

for any $c \in R$, $\overline{t}-r^2 \leq t_1 < t_2 \leq \overline{t}+r^2$; *then*

$$\sup_{(\overline{t}-\vartheta r^2,\overline{t}+\vartheta r^2)} \|v\|_{\vartheta r}^2 \leq \frac{Kr^{N+2}(1-\vartheta)^{-1}}{\Gamma\text{-cap}(N,Q(\overline{z},r))} \int_{Q(\overline{z},r)} |D_x v|^2 dxdt$$

$(\vartheta \in (0,1))$, *where* $N = \{z \in Q(\overline{z},\ r) \quad v(z) = 0\}$.

The proof of Lemma 1 is obtained proving a standard Poincaré's inequality for the case where the average of v on $Q(\overline{z},\vartheta r)$ is 0.

This latter estimate is obtained by contradiction, using the following compactness result for sequences of subsolutions of our problem.

LEMMA 2. *Let* $\{v_j\}$ *be a sequence of subsolutions of our problem in* $Q(\overline{z},r)$; *there exists a subsequence which is almost uniformly convergent in* $Q(\overline{z},r)$.

We give now the proof of Theorem 1.

Proof of Theorem 1. Let $z_0 \in \Sigma$, $k < 1 = f(z_0) = g(z_0) = u(z_0)$ weakly; there is $R_0 > 0$ such that $(u-k)^- \rightharpoonup 0$ weakly in $Q(z_0,R_0) \cap \Sigma$.

We define G^z_ρ (the regularized Green function of our problem) by

$$\int_{T_1}^{T_2} \int_B \{(D_t \varphi)\, G^z_\rho + \Sigma^N_{i,j=1} a_{ij} D_{x_j} G^z_\rho D_{x_i} \varphi\} dxdt =$$

$$= |Q(z,\rho)| \int_{Q(z,\rho)} \varphi \, dxdt$$

for any $\varphi \in C^\infty(\hat{Q})$, $\varphi(x,T_1) = 0$, $\varphi = 0$ on $\partial B \times (T_1,T_2)$, where $Q \subset\subset \hat{Q} = B \times (T_1,T_2)$ with B ball in R^N (the functions a_{ij} are defined as δ_{ij} in $\hat{Q} - Q$).

We use as test function in our problem $(u-k)^- G^z \eta^2 \tau^2$, where $Q(\overline{z},R) \subset Q(z_0,R_0)$ and $\eta = \eta(x)$, $\tau = \tau(t)$ are cutoff functions such that

$$\eta = 1 \text{ for } x \in B(\overline{x},R/8), \quad \eta = 0 \quad \text{for} \quad x \notin B(\overline{x},R/4);$$

$$0 \le \eta \le 1, \quad \eta \in C^\infty_0(R^N), \quad |D_x \eta| \le CR^{-2};$$

$$\tau = 1 \text{ for } t \ge \overline{t}-3\vartheta R^2, \ \tau = 0 \quad \text{for } t \le \overline{t} - (1-2\vartheta)R^2;$$

$$0 \le \tau \le 1, \quad \tau \in C^\infty_0(R), \quad |D_t \tau| \le CR^{-2}.$$

By standard methods and going to the limit as $\rho \to 0$ we obtain

$$\int_Q |D_x(u-k)^-|^2 G^{\overline{z}} \eta^2 \tau^2 \ dxdt + |(u-k)^-|^2(\overline{z}) \le \tag{2.2}$$

$$\le C\, [R^{-2} \int_{\overline{t}-(1-2\vartheta)R^2}^{\overline{t}} \int_{B(\overline{x},R/4)-B(\overline{x},R/8)} |(u-k)^-|^2 G^{\overline{z}} \ dxdt \ +$$

$$+ \int_{\overline{t}-(1-2\vartheta)R^2}^{\overline{t}-3\vartheta R^2} \int_{B(\overline{x},R/4)} |(u-k)^-|^2 G^{\overline{z}} \ dxdt \ +$$

$$+ \int_{\bar{t}-(1-2\vartheta)R^2}^{\bar{t}} \int_{B(\bar{x},R/4)} |(u-k)^-|^2 \tau^2 \eta |D_x\eta| \; |D_x G^{\bar{z}}| \; dxdt],$$

where $G^{\bar{z}}$ is the Green function of our problem with singularity at $\bar{z}$. We consider now the last term in the right-hand side; we have

$$\int_{\bar{t}-(1-2\vartheta)R^2}^{\bar{t}} \int_{B(\bar{x},R/4)} |(u-k)^-|^2 \tau^2 \eta |D_x\eta| \; |D_x G^{\bar{z}}| \; dxdt \le \tag{2.3}$$

$$\le \varepsilon R^{-N/2} \int_{\bar{t}-(1-2\vartheta)R^2}^{\bar{t}} \int_{B(\bar{x},R/4)-B(\bar{x},R/8)} \eta^2\tau^2 |(u-k)^-|^2 |D_x G^{\bar{z}}| |G^{\bar{z}}|^{-3/2} dxdt$$

$$+ 4/\varepsilon \; R^{(N/2)-2} \int_{\bar{t}-(1-2\vartheta)R^2}^{\bar{t}} \int_{B(\bar{x},R/4)-B(\bar{x},R/8)} |(u-k)^-|^2 |G^{\bar{z}}|^{3/2} \; dxdt,$$

where $\varepsilon > 0$ is to be chosen.

Since u is a solution of our problem, we obtain, after some calculations,

$$\int_{\bar{t}-(1-2\vartheta)R^2}^{\bar{t}} \int_{B(\bar{x},R/4)-B(\bar{x},R/8)} \eta^2\tau^2 |(u-k)^-|^2 |D_x G^{\bar{z}}| |G^{\bar{z}}|^{-3/2} dxdt \le \tag{2.4}$$

$$\le C \; [R^{-2} \int_{\bar{t}-(1-2\vartheta)R^2}^{\bar{t}} \int_{B(\bar{x},R/4)-B(\bar{x},R/16)} |(u-k)^-|^2 \tau^2 (|G^{\bar{z}}|^{\frac{1}{2}} + R^{-N/2}) dxdt \; +$$

$$+ R^{N/2} \int_{\bar{t}-(1-2\vartheta)R^2}^{\bar{t}} \int_{B(\bar{x},R/4)} |D_x(u-k)^-|^2 \; G^{\bar{z}} \; \eta^2 \; \tau^2 \; dxdt].$$

From (2.2), (2.3), (2.4), choosing a suitable ε we obtain

$$\int_{\bar{t}-(1-2\vartheta)R^2}^{\bar{t}} \int_{B(\bar{x},R)} |D_x(u-k)^-|^2 \; \eta^2\tau^2 \; G^{\bar{z}} \; dxdt + |(u-k)^-|^2(\bar{z}) \le \tag{2.5}$$

$$\le C \; [\int_{\bar{t}-(1-2\vartheta)R^2}^{\bar{t}} \int_{B(\bar{x},R/4)-B(\bar{x},R/16)} |(u-k)^-|^2 \{G^{\bar{z}} + R^{-N/2}(G^{\bar{z}})^{1/2} \; +$$

$$+ R^{N/2}(G^{\bar{z}})^{3/2} + R^{-N}\} \; dxdt + \int_{\bar{t}-(1-2\vartheta)R^2}^{\bar{t}} \int_{B(\bar{x},R/4)} |(u-k)^-|^2 \{G^{\bar{z}} \; +$$

$$+ R^{-N/2}(G^{\bar{z}})^{1/2} + R^{N/2}(G^{\bar{z}})^{3/2} + R^{-N}\} \; dxdt].$$

Using well-known estimates on the Green function [1], we obtain:

$$\int_{\overline{t}-R^2}^{\overline{t}} \int_{B(\overline{x},\vartheta^{\frac{1}{2}}R)} |D_x(u-k)^-|^2 G^{\overline{z}}\, dxdt + |(u-k)^-|^2(\overline{z}) \leq$$

$$\leq C \exp(-C\vartheta^{-1})\vartheta^{-3N/4} \sup_{B(\overline{x},R/4)\times(\overline{t}-3\vartheta R^2,\overline{t})} |(u-k)^-|^2 +$$

$$+ C\vartheta^{-3N/4} R^{-(N+2)} \int_{\overline{t}-(1-2\vartheta)R^2}^{\overline{t}-3\vartheta R^2} \int_{B(\overline{x},R/4)} |(u-k)^-|^2\, dxdt$$

(in the following we denote by C any constant depending only on N,λ,Λ).

Taking the supremum for $\overline{z} \in Q(z_0,\vartheta^{\frac{1}{2}}R)$ we obtain the following "Caccioppoli's inequality":

$$\int_{Q(z_0,\vartheta^{\frac{1}{2}}R)} |D_x(u-k)^-|^2 G^{z_0} dxdt + \sup_{Q(z_0,\vartheta^{\frac{1}{2}}R)} |(u-k)^-|^2 \leq \tag{2.6}$$

$$\leq C \exp(-C\vartheta^{-1})\, \vartheta^{-3N/4} \sup_{Q(z_0,R)} |(u-k)^-|^2 +$$

$$+ C\vartheta^{-3N/4} R^{-(N+2)} \int_{t_0-(1-\vartheta)R^2}^{t_0-2\vartheta R^2} \int_{B(x_0,R/2)} |(u-k)^-|^2\, dxdt \leq$$

$$\leq CK_1(\vartheta) \sup_{Q(z_0,R)} |(u-k)^-|^2 +$$

$$+ C(K_1(\vartheta)\delta_\vartheta(R))^{-1} \int_{t_0-R^2}^{t_0-R^2} \int_{B(x_0,R)} |D_x(u-k)^-|^2\, G^{z_0}\, dxdt,$$

where $K_1(\vartheta) = \exp(-C\vartheta^{-1})\vartheta^{-3N/4}$ and use of the Poincaré inequality has been made. From (2.6) denoting

$$\mu(r) = \sup_{Q(z_0,r)} |(u-k)^-|^2 ,$$

$$\nu(r) = \int_{Q(z_0,r)} |D_x(u-k)^-|^2\, G^{z_0}\, dxdt,$$

and choosing $\vartheta \in (0,\vartheta_0)$ with a suitable ϑ_0, we obtain after some calculations:

$$\mu(\vartheta^{\frac{1}{2}}R) + \nu(\vartheta^{\frac{1}{2}}R) \leq (1+CK_1(\vartheta))^{-1} \quad (\mu(R) + \nu(R)). \tag{2.7}$$

From (2.7), using the integration lemma in [12] (see also [7]), we obtain

$$\mu(r) + \nu(r) \leq \exp(-\beta\int_r^{\overline{R}} \delta_\vartheta(\rho)\, d\rho/\rho)\ (\mu(\overline{R}) + \nu(\overline{R})) \tag{2.8}$$

$(\overline{R} \leq \vartheta R_o)$. Choosing now $\overline{R} = \omega_\vartheta(r)^\alpha$ we obtain easily

$$\mu(r) + \nu(r) \leq \omega_\vartheta(r)^{\beta(1-\alpha)} + \Psi_f(\omega_\vartheta(r)^\alpha);$$

then

$$\sup_{Q(z_o,r)} (u(z)-u(z_o))^- \leq K\omega_\vartheta(r)^{\beta(1-\alpha)} + \Psi_f(\omega_\vartheta(r)^\alpha). \tag{2.9}$$

An analogous estimate holds also for $(u(z)-u(z_o))^+$; from both estimates the result follows.

REFERENCES

1. Aronson, D.G.: Nonnegative solutions of linear parabolic equations, Ann. Scuola Norm. Sup. Pisa 22 (1968), 607-694.
2. Biroli, M. and U. Mosco: Estimations ponctuelles sur le bord du module de continuité des solutions faibles des équations paraboliques, C.R. Acad Sci. Paris 296 (1983), 841-843.
3. Biroli, M. and U. Mosco: Wiener estimates at boundary points for parabolic equations, Ann. Math. Pura Appl. (to appear).
4. Biroli, M. and U. Mosco: Wiener estimates for parabolic obstacle problems (I.M.A. preprint, University of Minnesota, 1985).
5. Biroli, M., U. Mosco and E. Di Benedetto: Wiener estimates for a class of free boundary problems (I.M.A. preprint, University of Minnesota, 1985).
6. Evans, L.C. and R.F. Gariepy: Wiener's criterion for the heat equation, Arch. Rational Mech. Anal. 72 (1982), 293-311.
7. Frehse, J. and U. Mosco: Wiener obstacles. In "Collège de France Seminar on nonlinear partial differential equations 1982/83" (Pitman, 1984).
8. Gariepy, R.F. and W.P. Ziemer: A regularity condition at the boundary for quasilinear elliptic equations, Arch. Rational Mech. Anal. 67 (1977), 25-39.
9. Gariepy, R.F. and W.P. Ziemer: Thermal capacity and boundary regularity, J. Differential Equations 45 (1982), 347-388.
10. Lanconelli, E.: Sul problema di Dirichlet per l'equazione del calore, Ann. Mat. Pura Appl. 106 (1975), 11-38.
11. Maz'ja, V.G.: Behaviour near the boundary of solutions of the Dirichlet problem for a second order elliptic equation in divergence form, Math. Notes 2 (1967), 610-617.

12. Maz'ja, V.G.: On the continuity at boundary points of the solution of quasilinear elliptic equations, Vestnik Leningrad Univ. Math. 3 (1976), 225-242.

13. Mosco, U.: Pointwise estimates for elliptic obstacle problem. In "Proceedings Berkeley Symposium on Nonlinear Functional Analysis 1983" (to appear).

14. Mosco, U.: Wiener potential estimates for the obstacle problem (I.M.A. preprint, University of Minnesota, 1985).

15. Ziemer, W.P.: Behaviour at the boundary of solutions of quasilinear parabolic equations, J. Differential Equations 35 (1980), 291-305.

M. Biroli
Dipartimento di Matematica
Politecnico di Milano
Via Bonardi 9
20133 Milano
ITALY

G CAGINALP & P C FIFE

Elliptic problems with layers representing phase interfaces

This is a summary of recent results [1] on steady solutions in two space dimensions of the "phase field model" for phase transitions. The simplest, and most commonly used, continuum model for a two-phase heat conducting material is the classical Stefan problem. The latter is inadequate in that it does not account for the phenomena of supercooling and superheating, the effects of the interface's curvature and surface tension on the equilibrium temperature at the interface, nor the finite thickness of the interface. All of these are physical realities, being manifestations of the finite correlation length between molecules in the material.

The phase field model was designed as a generalization of the Stefan model to account for those effects. In that model (see [2] and references therein), which is based on a Landau-Ginzburg concept of free energy, the phase of the material is given by the sign of a continuous "order parameter" function $\varphi(x,t)$. The latter is coupled to the temperature by a pair of nonlinear elliptic equations. Here, we are concerned with the steady state version, which takes the form

$$\xi^2 \Delta\varphi + g(\varphi) + 2u = 0 \quad \text{in} \quad \Omega \tag{1}$$
(a bounded smooth domain in R^2),

$$\Delta u = 0 \quad \text{in} \quad \Omega,$$

plus boundary conditions on $\partial\Omega$. Here Ω represents the region occupied by the material. The phase is determined by φ. When $\varphi > 0$, the material is in the less ordered state (liquid, say), and when $\varphi < 0$, it is in the other state (solid, say). Boundary conditions for u (taken for definiteness to be of Dirichlet or Robin type) will determine u uniquely, of course, so in the steady-state situation, the problem reduces to a single equation (1) with a known inhomogenity $u(x)$. The parameter $\xi \ll 1$ is the correlation length

Research supported by NSF Grants DMS8403184 and DMS8503007.

referred to above. The function g is here taken to be a standard bistable cubic

$$g(\varphi) = \frac{1}{2}(\varphi - \varphi^3),$$

so that $\varphi = \pm 1$ are stable rest states of the equation $y'(t) = g(y(t))$.

The first result is intuitively clear but nontrivial to prove.

THEOREM 1. *Suppose the nullset* $\{x : u(x) = 0\}$ *is a simple closed curve* $\Gamma_0 \in \Omega$ *on which* $\nabla u \neq 0$. *Suppose* $u > 0$ *in the exterior of* Γ_0 *(which we call* Ω_+*), and* $u < 0$ *in the interior* (Ω_-). *Then for each* $\xi > 0$, (1) *has a solution* φ_ξ, *satisfying zero Neumann conditions on* $\partial\Omega$, *such that*

$$\lim_{\xi \to 0} \varphi_\xi(x) = \pm 1$$

uniformly for x *in closed subsets of* Ω.

No uniqueness is implied; just the existence of a family φ_ξ with that property. It is a layered family in the sense that for small ξ, a transition between 1 and -1 occurs near the interface between Ω_+ and Ω_-. The same result is true for suitable Dirichlet conditions; in fact a proof for that case was given in [3]. That same proof works for Neumann conditions; alternatively, one could proceed by upper and lower solutions. The proof in [3] yields much more detail about the nature of the transition than is stated in the theorem.

This result does not bring out the effects of curvature and surface tension, as referred to above, until one makes a closer analysis of the transition region. These effects are most evident when the temperature is $O(\xi)$, which is assumed from now on. So let $2u(x) \equiv \xi\tilde{u}(x)$; (1) now takes the form

$$\xi^2 \Delta\varphi + g(\varphi) + \xi\tilde{u} = 0 \quad \text{in} \quad \Omega \in R^2. \tag{2}$$

For definiteness, impose Neumann conditions

$$\frac{\partial\varphi}{\partial\eta} = 0 \quad \text{on} \quad \partial\Omega \,. \tag{3}$$

Formal asymptotics, which will not be given here, imply that if a solution of (2), (3) with internal layer along a curve Γ exists, then the temperature at points on the layer is related linearly to its curvature:

$$\tilde{u}(x) = -\gamma k(x) \quad \text{on} \quad \Gamma, \tag{4}$$

where k is the curvature of Γ and γ is a specific constant:

$$\gamma = \frac{1}{2}\int_{-\infty}^{\infty}\left(\frac{\partial\chi(\rho)}{\partial\rho}\right)^2 d\rho,$$

$$\chi(\rho) = \tanh\frac{\rho}{2}\,.$$

This result is in agreement with physics; in fact the surface tension effect is given by the Gibbs-Thompson relation

$$u(x) = -\frac{\sigma}{\Delta S}\,k(x) \quad \text{on an interface,} \tag{5}$$

where σ represents surface tension and ΔS is the entropy difference involved in the phase change. Next, it is noted that σ is related (linearly, to first order) to the correlation length ξ, so that it is appropriate to write $\sigma = \xi\sigma^*$, $\sigma^* = O(1)$. It then follows that (5) is the same as (4), provided

$$\gamma = \frac{2\sigma^*}{\Delta S}\,.$$

The main result to be reported here is the existence of layered and unlayered solutions of (2), (3), the layered ones having layer at a curve Γ satisfying (4).

As a first step, the existence of such a closed curve Γ will be established under certain conditions. To formulate them, we use polar coordinates

$$(r,\varphi) \quad \text{in} \quad \Omega \text{ and set} \quad w = \frac{1}{r}, \quad F(w,\vartheta) = -\gamma\tilde{u}\left(\frac{1}{w},\vartheta\right)\,.$$

<u>THEOREM 2.</u> *Assume there exist positive numbers α and β such that the circles $\{w = \alpha\}$ and $\{w = \beta\}$ both lie in Ω, as well as the annulus between them, and such that*

$$\beta - F(\beta,\vartheta) < 0 < \alpha - F(\alpha,\vartheta) \quad \text{for all} \quad \vartheta \in [-\pi,\pi]$$

and

$$0 < \beta - \alpha < K,$$

where K is a certain specific number depending only on F. Then there exists a simple closed curve $\Gamma \subset \Omega$ on which (4) holds.

The proof is by reduction of the problem to a periodic problem for a nonlinear second order differential equation (with cubic nonlinearity in the first derivative) and using the method of upper and lower solutions to prove existence. (The condition involving K is to ensure a Nagumo-type condition for the cubic non-linearity.)

The next result assumes the existence of a simple closed curve $\Gamma \subset \Omega$ on which (4) holds, and on which $\nabla\tilde{u} \neq 0$. It may or may not be the one ensured by Theorem 2. It divides Ω into two parts, Ω_+ and Ω_-, defined in the following way. The normal derivative of $\tilde{u}$ on Γ is nonzero, so is positive in the direction to one side of Γ or the other. Choose Ω_+ to be on the positive side, so near Γ in Ω_+, $\tilde{u}$ is greater than $\tilde{u}$ on Γ itself.

<u>THEOREM 3.</u> *Under the above assumption, there exist, for each small enough ξ, at least three solutions φ_ξ^+, φ_ξ^-, φ_ξ^o of (2), (3). They satisfy*

$$\lim_{\xi\to 0} \varphi_\xi^\pm = \pm 1 \tag{6}$$

uniformly for $x \in \Omega$. Also φ_ξ^o satisfies (6) with the + sign uniformly in closed subsets of Ω_+, and (6) with the - sign in closed subsets of Ω_-.

Thus, φ_ξ^o is a layered solution with layer near Γ. It represents a material which is in one phase (solid, say) in Ω_-, and another (liquid) in Ω_+. The interface between the two phases indeed satisfies the Gibbs-Thompson relation, which in this formulation arises automatically from the mathematics, rather than an additional physical constraint. The physics embodied by the Landau-Ginzburg formulation is thus sufficient to guarantee this relation.

The other two solutions φ_ξ^+ and φ_ξ^- represent states which are purely solid or purely liquid. In part of the region Ω (namely where $\tilde{u} < 0$), the material in state φ_ξ^+ will be a supercooled liquid. The existence of a solution φ_ξ^+ is consistent with the observation in crystal growth that one generally requires a "seed" to initiate growth of the solid. This requirement is often removed when the temperature is so low that the Gibbs-Thompson relation would imply a radius of curvature which is comparable to a "cluster" of liquid atoms [6]. A similar superheating effect holds.

Much more detailed information about the closeness of the approximation of φ_ξ^o, φ_ξ^+, and φ_ξ^- to their respective limits for small ξ is given in [1], as well as information about the width of the transition layer.

A similar theorem is available for suitable Dirichlet problems, except in that case we do not have the multiplicity; the type of solution obtained (layered, all solid, or all liquid) depends on the boundary conditions.

Related results for radially symmetric problems in arbitrary dimensions are given in [4] and [5].

REFERENCES

1. Caginalp, G. and P.C. Fife: Elliptic problems involving phase boundaries satisfying a curvature condition (preprint, 1985).

2. Caginalp, G.: An analysis of a phase field model of a free boundary, Arch. Rat. Mech. Anal. (to appear).

3. Fife, P.C. and W.M. Greenlee: Interior transition layers for elliptic boundary value problems with a small parameter, Russ. Math. Surveys 29 (1974), 103-131.

4. Caginalp, G. and S. Hastings: Properties of some ordinary differential equations related to free boundary problems (Univ. of Pittsburgh preprint, 1984).

5. Caginalp, G. and B. McLeod: The interior transition layer for an ordinary differential equation arising from solidification theory (Univ. of Pittsburgh, preprint, 1985).

6. Chalmers, B.: Principles of Solidification (Krieger, 1977).

G. Caginalp
Mathematics Department
University of Pittsburgh
Pittsburgh, PA 15260
U.S.A.

P.C. Fife
Mathematics Department
University of Arizona
Tucson, AZ 85721
U.S.A.

S CAMPANATO

$L^{2,\lambda}$ theory and nonlinear parabolic systems

We shall denote by Ω a bounded open set in R^n, $n \geq 1$, $x = (x_1,\ldots,x_n)$ a point in Ω, by $\mathcal{C}$ the cylinder $\Omega \times (-T,0)$, $T > 0$, $X = (x,t)$ a point in $\mathcal{C}$, by $u(X)$ a vector $\mathcal{C} \to R^N$, $N > 1$, and

$$u' = \frac{\partial u}{\partial t},\ D_i u = \frac{\partial u}{\partial x_i}\ ,\quad U = Du = (D_1 u,\ldots,D_n u)\ .$$

We provide R_X^{n+1} with the parabolic metric: if $X = (x,t)$ and $Y = (y,\tau)$ then

$$\delta(X,Y) = \max\{\|x-y\|,\ |t-\tau|^{\frac{1}{2}}\}.$$

We shall set, for $X_o = (x^o,\ t_o)$,

$$B(x^o,\sigma) = \{x \in R^n\colon \|x-x^o\| < \sigma\}$$
$$Q(X_o,\sigma) = \{X\colon\delta(X,X_o) < \sigma \text{ and } t < t_o\} = B(x^o,\sigma) \times (t_o-\sigma^2,t_o)$$

and we shall say that $Q(X_o,\sigma) \subset\subset \mathcal{C}$ if $B(x^o,\sigma) \subset\subset \Omega$ and $\sigma^2 < t_o + T \leq T$.

We consider the differential system

$$\frac{\partial u}{\partial t} = \sum_i D_i a^i(Du) = \overline{E}_o(u). \tag{1}$$

It is not restrictive to assume $a^i(0) = 0$. $\overline{E}_o$ is a strongly elliptic operator in x with a non linearity 2. This means that $p \to a^i(p)$ are vectors on R^N of class $C^1(R^{nN})$ and there exist constants $\nu > 0$ and $M > 0$ such that, $\forall\ p = (p^1,\ldots,p^n)\ R^{nN}$ and $\xi = (\xi^1,\ldots,\xi^n)\ R^{nN}$,

$$\nu\|\xi\|^2 \leq \sum_{ij} (A_{ij}(p)\xi^j|\xi^i) \leq M\|\xi\|^2 \tag{2}$$

where we have set

$$A_{ij}(p) = \{A^{hk}_{ij}(p)\}, \quad A^{hk}_{ij} = \frac{\partial a^i_h}{\partial p^j_k}. \tag{3}$$

Let us set

$$u_A = \fint_A u(X)dX$$

and

$$\Phi(u,X_o,\sigma) = \int_{Q(X_o,\sigma)} \|Du\|^2 + \sigma^{-2}\|u-u_{Q(\sigma)}\|^2 \, dX. \tag{4}$$

Let $u \in L^2(-T,0;H^1(\Omega))$ be a solution of the system (1) in the usual weak sense:

$$\int_C \sum_i (a^i(Du)|D_i\varphi) - (u|\varphi')dX = 0, \qquad \forall\varphi \in C^\infty_o(C). \tag{5}$$

The following sequence of propositions can be proved:

(I) $\forall Q(2\sigma) \subset C$

$$D_{ij}u \in L^2(Q(\sigma)), \quad ij = 1,\dots,n \tag{6}$$

$$u' \in L^2(Q(\sigma)) \tag{7}$$

and we have the following estimate (of Caccioppoli type)

$$\int_{Q(\sigma)} \|DU\|^2 + \|u'\|^2 dX \le \frac{c(\nu,M)}{\sigma^2} \int_{Q(2\sigma)} \|U\|^2 \, dX \tag{8}$$

(see [3] Theorem 3.1). As a consequence we have (see [3] Theorem 4.1)

(II) $\forall Q(2\sigma) \subset C$ *we have the estimates (of Poincaré type)*

$$\Phi(u,X_o,\sigma) \le c(\nu,M) \int_{Q(2\sigma)} \|Du\|^2 \, dX \tag{9}$$

$$\Phi(U,X_o,\sigma) \le c(\nu,M) \int_{Q(\sigma)} \|DU\|^2 \, dX. \tag{10}$$

It can easily be seen that, in each $Q(\sigma) \subset\subset \mathcal{C}$, the vector $U = Du$ is a (weak) solution of the quasi-linear system

$$\frac{\partial U}{\partial t} = \sum_{ij} D_i \{\vartheta_{ij}(U) D_j U\} , \tag{11}$$

where

$$\vartheta_{ij}(p) = \left\{ \begin{array}{c|c|c} A_{ij}(p) & & \\ \hline & \ddots & \\ \hline & & A_{ij}(p) \end{array} \right\} , \; n^2 \; N \times N \text{ blocks} .$$

The system (11) can be considered as a strongly parabolic linear system with coefficients $\vartheta_{ij}(U(X)) \in L^\infty(Q(\sigma))$ and hence (see [3] Theorem 5.1)

(III) $\exists \varepsilon \in (0,n+2]$ *such that,* $\forall Q(X_o,\sigma) \subset\subset \mathcal{C}$ *and* $\forall \lambda \in (0,1)$,

$$\int_{Q(\lambda\sigma)} \|DU\|^2 dX \le c\lambda^\varepsilon \int_{Q(\sigma)} \|DU\|^2 dX \tag{12}$$

where c *does not depend on* X_o, λ *and* σ.

We observe that $\varepsilon = n+2$ if (1) is a basic linear system (see [2]) but, in general, ε can be aribtrarily near zero. This is a difference from the case $N = 1$ in which, by a classical result of De Giorgi and Nash, U is Hölder continuous in $\mathcal{C}$ and hence $\varepsilon > n$.

(IV) It follows from (12) that:
If u *is a (weak) solution of the system* (1) *in* $\mathcal{C}$ *then,* $\forall Q(\sigma) \subset\subset \mathcal{C}$ *and* $\forall \lambda \in (0,1)$,

$$\int_{Q(\lambda\sigma)} \|Du\|^2 dX \le c\lambda^{\varepsilon_o} \int_{Q(\sigma)} \|Du\|^2 dX \tag{13}$$

where, $\forall \mu \in (0,n)$,

$$\varepsilon_o = \min\{\varepsilon,n\} + 2 - \mu\delta_{\varepsilon,n} \tag{14}$$

In fact, it follows from (12) and by the Poincaré inequality (10), that, if $0 < \tau < \frac{1}{2}$,

$$\int_{Q(\tau\sigma)} \|U-U_{Q(\tau\sigma)}\|^2 dX \le c(\tau\sigma)^2 \int_{Q(\tau\sigma)} \|DU\|^2 dX \le$$

$$\le c(\nu,M)\sigma^2\tau^{2+\varepsilon} \int_{Q(\sigma/2)} \|DU\|^2 dX.$$

On the other hand, if $0 < \lambda < \tau < \frac{1}{2}$, we have

$$\int_{Q(\lambda\sigma)} \|U\|^2 dX \le c\left(\frac{\lambda}{\tau}\right)^{n+2} \int_{Q(\tau\sigma)} \|U\|^2 dX + c \int_{Q(\tau\sigma)} \|U-U_{Q(\tau\sigma)}\|^2 dX.$$

Finally, if $0 < \lambda < \tau < \frac{1}{2}$, we have

$$\int_{Q(\lambda\sigma)} \|U\|^2 dX \le c\left(\frac{\lambda}{\tau}\right)^{n+2} \int_{Q(\tau\sigma)} \|U\|^2 dX + c\sigma^2\tau^{2+\varepsilon} \int_{Q(\sigma/2)} \|DU\|^2 dX. \tag{15}$$

Then, if $\varepsilon < n$, it follows, by Lemma 1.1 page 7 of [1] and from (15), that

$$\int_{Q(\lambda\sigma)} \|U\|^2 dX \le c\left(\frac{\lambda}{\tau}\right)^{2+\varepsilon} \int_{Q(\tau\sigma)} \|U\|^2 dX + c\sigma^2\lambda^{2+\varepsilon} \int_{Q(\sigma/2)} \|DU\|^2 dX.$$

The estimate (13) follows from this on passing to the limit as $\tau \to \frac{1}{2}$ and recalling the estimate of Caccioppoli (8).

Finally, we note that (13) is trivially true also for $\frac{1}{2} \le \lambda < 1$.

If instead $n < \varepsilon \le n+2$ it follows, from (12), the inequalities of Poincaré (10) and of Caccioppoli (8), that $\forall Q(\sigma) \subset\subset \mathcal{C}$ and $\forall \lambda \in (0,1)$

$$\int_{Q(\lambda\sigma)} \|U-U_{Q(\lambda\sigma)}\|^2 dX \le c\lambda^{2+\varepsilon} \int_{Q(\sigma)} \|U\|^2 dX .$$

Hence $U \in C^{0,\alpha}(\mathcal{C},\delta)$, $\alpha = \frac{\varepsilon-n}{2}$, and $\forall Q(\sigma) \subset\subset \mathcal{C}$

$$\sigma^{2+\varepsilon}[U]^2_{\alpha,\overline{Q(\sigma/2)}} \le c \int_{Q(\sigma)} \|U\|^2 dX . \tag{16}$$

On the other hand, $\forall X,Y \in Q(\sigma/2)$, we have

$$\|U(X)\|^2 \le 2\|U(Y)\|^2 + c\sigma^{2\alpha}[U]^2_{\alpha,\overline{Q(\sigma/2)}} . \tag{17}$$

Finally, we have the estimate

$$\sigma^{n+2} \sup_{Q(\sigma/2)} \|U(X)\|^2 \le c \int_{Q(\sigma)} \|U\|^2 dX. \tag{18}$$

Hence, $\forall 0 < \lambda < \frac{1}{2}$, we have

$$\int_{Q(\lambda\sigma)} \|U(X)\|^2 dX \leq c(\lambda\sigma)^{n+2} \sup_{Q(\sigma/2)} \|U(X)\|^2 \leq c\lambda^{n+2} \int_{Q(\sigma)} \|U\|^2 dX \ . \qquad (19)$$

This estimate is trivially true also for $\frac{1}{2} \leq \lambda < 1$.

Finally, if $\varepsilon = n$, the estimate (12) is true also for $\varepsilon = n-\mu$, $\forall \mu \in (0,n)$. Hence, repeating the argument used in the case of $\varepsilon < n$, we again obtain the estimate (13) with ε_0 defined as in (14).

In particular, it follows from the estimates of Poincaré (9) and of Caccioppoli ([3] Lemma 2.X) and from (13), that

$$\int_{Q(\lambda\sigma)} \|u-u_{Q(\lambda\sigma)}\|^2 dX \leq c(\nu,M)\lambda^{2+\varepsilon_0} \int_{Q(\sigma)} \|u\|^2 dX \ .$$

Hence, if $\varepsilon_0 < n$ (in particular if $n \leq 2$),

$$u \in C^{0,\alpha}(\mathcal{C},\delta) \quad \text{with} \quad \alpha = \frac{\varepsilon_0 - n}{2} \qquad (20)$$

and $\forall Q(2\sigma) \subset \mathcal{C}$

$$\sigma^{2+\varepsilon_0}[u]^2_{\alpha,\overline{Q(\sigma)}} \leq c \int_{Q(2\sigma)} \|u\|^2 dX \qquad (21)$$

$$\sigma^{n+2} \sup_{Q(\sigma)} \|u\|^2 \leq c \int_{Q(2\sigma)} \|u\|^2 dX \ . \qquad (22)$$

(V) In the same way as (13) follows from (12), using an analogous argument, we obtain from (13) that, *if* u *is a (weak) solution of the system* (1) *in* $\mathcal{C}$ *then,* $\forall Q(\sigma) \subset\subset \mathcal{C}$ *and* $\forall \lambda \in (0,1)$, *we have*

$$\int_{Q(\lambda\sigma)} \|u\|^2 dX \leq c\lambda^{\varepsilon_1} \int_{Q(\sigma)} \|u\|^2 dX \qquad (23)$$

where, $\forall \mu \in (0,n)$,

$$\varepsilon_1 = \min\{\varepsilon_0, n\} - \mu\delta_{\varepsilon_0, n} \ . \qquad (24)$$

In particular, if $\varepsilon_0 > n$ (and surely if $n \leq 2$), then $\varepsilon_1 = n+2$ and

$$\int_{Q(\lambda\sigma)} \|u\|^2 dX \leq c \int_{Q(\sigma)} \|u\|^2 dX \; . \tag{25}$$

We wish to indicate some important consequences of the propositions (I),...,(V).

We shall denote by $\omega(\sigma)$ a function, defined for $\sigma \geq 0$, having the following properties: $\omega(\sigma)$ is non negative, non decreasing, bounded, continuous, concave and $\omega(0) = 0$.

We recall that a vector $v : \mathcal{C} \to R^N$ is partially α-Hölder continuous in $\mathcal{C}$ if $\exists$ a subset $\mathcal{C}_o \subset \mathcal{C}$ ($\mathcal{C}_o$ is the singular set of the vector v) such that

$\mathcal{C}_o$ is closed in $\mathcal{C}$

meas $\mathcal{C}_o = 0$

$v \in C^{o,\alpha}(\mathcal{C}\ \mathcal{C}_o, \delta)$.

$H_\beta(\mathcal{C}_o)$ is the β-dimensional Hausdorff measure of $\mathcal{C}_o$, Hausdorff measure with respect to the parabolic metric δ.

a) If $u \in L^2(-T,0;H^1(\Omega))$ is a solution of the basic system (1) and if, $\forall p,\bar{p} \in R^{nN}$,

$$\{\sum_{ij} \|A_{ij}(p) - A_{ij}(\bar{p})\|^2\}^{\frac{1}{2}} \leq \omega(\|p-\bar{p}\|^2) \tag{27}$$

then Du is partially α-Hölder continuous in $\mathcal{C}$, $\forall\alpha \in (0,1)$, and if $\mathcal{C}_o$ is the singular set of the vector Du then we have

$$H_{n+2(1-t_o)}(\mathcal{C}_o) = 0, \quad \text{for some } t_o > 1. \tag{28}$$

In fact, U = Du is a solution, at least locally in $\mathcal{C}$, of the quasi-linear system (11). The result then follows as shown in [6] and [4].

b) Let us consider the non linear system

$$\frac{\partial u}{\partial t} = E_o(u) + F(u) \tag{29}$$

where $E_o(u) = \sum_i D_i a^i(X,u,Du)$ is a strongly elliptic operator in x, with a non-linearity 2, while $F(u) = -\sum_i D_i B^i(X,u) + B(X,u,Du)$ has a controlled growth in u and Du (see [3], section 1, page 86).

Let $u \in L^2(-T,0;H^1(\Omega)) \cap L^\infty(-T,0,L^2(\Omega))$ be a weak solution of the system (29). If $n < \varepsilon_0$ (in particular, if $n \leq 2$) and

$$\{\sum_i \|a^i(X,u,p)-a^i(Y,v,p)\|^2\}^{\frac{1}{2}} \leq \omega(\delta^2(X,Y) + \|u-v\|^2)\|p\|, \tag{30}$$

then u is partially α-Hölder continuous in $\mathcal{C}$, with $\alpha = \frac{\varepsilon_0 - n}{2}$, and we have

$$H_{n+2(1-t_0)}(\mathcal{C}_0) = 0, \quad \text{for a} \quad t_0 > 1. \tag{31}$$

If, in addition to the condition (30), the following condition

$$\sum_{ij} \|\tilde{A}_{ij}(X,u,p) - \tilde{A}_{ij}(X,u,\overline{p})\|^2 \leq (\|p-\overline{p}\|^2) \tag{32}$$

is also satisfied, then, for any arbitrary n,u is partially α-Hölder continuous in $\mathcal{C}$, $\forall\alpha < 1$; however, regarding the singular set $\mathcal{C}_0$ we can only say that meas $\mathcal{C}_0 = 0$.

It is already known in the literature that, if the system is quasi linear

$$a^i(X,u,p) = \sum_j \tilde{A}_{ij}(X,u)p^j \tag{33}$$

and the condition (30) holds, then, for any arbitrary n,u is partially α-Hölder continuous in $\mathcal{C}$, $\forall\alpha < 1$, and (31) also holds (see [6] for the case F = 0, [4] when F has a linear growth, [5] and [8] when F has a controlled growth).

c) Suppose now that $u \in L^2(-T,0;H^1(\Omega)) \cap L^\infty(\mathcal{C})$ is a solution of the system (29) and F(u) has natural growth, that is, if $K = \sup_{\mathcal{C}} \|u\|$,

$$\|B^i(X,u)\| \leq c(K)$$

$$\|B(X,u,Du)\| \leq c(K) + b(K)\ \|Du\|^2$$

$$2Kb(K) < \nu .$$

It is known that, if the system is quasilinear and the condition (30) holds then, for any n and $\forall\alpha < 1$,u is partially α-Hölder continuous in $\mathcal{C}$ and

$$H_{n+2(1-t_o)}(\mathcal{C}_o) = 0, \quad \text{for some } t_o > 1. \tag{34}$$

This result holds also for non linear systems (29) with the difference that, regarding the singular set $\mathcal{C}_o$ we can only say that meas $\mathcal{C}_o = 0$.

As a consequence of the proposition (V), one should be able to control the Hausdorff measure of the set $\mathcal{C}_o$ exactly as in (34), provided that $n < \varepsilon_o$ (in particular, if $n \leq 2$).

REFERENCES

1. Campanato, S.: Sistemi ellittici in forma divergenza. Regolarità all'interno, Quad. Scuola Norm. Sup. Pisa, 1980.
2. Campanato, S.: Equazioni paraboliche del secondo ordine e spazi $\mathcal{L}^{2,\vartheta}(\Omega,\delta)$, Ann. Mat. Pura Appl. 73 (1966), 55-102.
3. Campanato, S.: On the non-linear parabolic systems in divergence form. Hölder continuity and partial Hölder continuity of the solutions, Ann. Mat. Pura Appl. 137 (1984), 83-122.
4. Campanato, S.: Partial Hölder continuity of solutions of quasi-linear parabolic systems of second order with linear growth, Rend. Sem. Mat. Univ. Padova 64 (1981), 59-75.
5. Campanato, S.: L^p regularity and partial Hölder continuity for solutions of second order parabolic systems with strictly controlled growth, Ann. Mat. Pura Appl. 128 (1980), 287-316.
6. Giaquinta, M. and E. Giusti: Partial regularity for the solutions to non-linear parabolic systems, Ann. Mat. Pura Appl. 97 (1973), 253-366.
7. Ladyzenskaya, O.A., V.A. Solonnikov and N.N. Ural'ceva: Linear and quasi-linear equations of parabolic type (Amer. Math. Soc., 1968).
8. Marino, M. and A. Maugeri: L^p theory and partial Hölder continuity for quasi-linear parabolic systems of higher order with strictly controlled growth. Ann. Mat. Pura Appl. (to appear).

S. Campanato
Dipartimento di Matematica
Università di Pisa
Via Buonarroti, 2
56100 Pisa
Italia

F CHIARENZA

Harnack inequality and degenerate parabolic equations

The aim of this paper is to review some recent aspects of the regularity theory for linear degenerate parabolic equations. As will be shown by the examples below, a certain kind of "bad" behaviour of the solutions to linear degenerate parabolic equations is reminiscent of analogous situations arising in the non linear case.

Let us consider in R^n an open bounded set Ω and, given $T > 0$, the correspondent cylinder $Q = \Omega \times]0,T[$ in R^{n+1}. In Q we will consider the equation

$$-\frac{\partial}{\partial x_i}\left(a_{ij}(x,t)\frac{\partial u}{\partial x_j}\right) + \frac{\partial u}{\partial t} = f \qquad (1)$$

where $a_{ij}(x,t) = a_{ji}(x,t)$ $(i,j = 1,\dots,n)$ are measurable functions in Q fulfilling the following condition

$$\exists\lambda > 0 : \frac{1}{\lambda} w(x,t)|\xi|^2 \leq a_{ij}(x,t)\xi_i\xi_j \leq \lambda w(x,t)|\xi|^2,$$

$$\forall\xi \in R^n, \quad \text{a.e. in } Q.$$

Here $w(x,t)$ is a nonnegative measurable function in Q.

In recent years the associate elliptic equation has been studied by Fabes, Kenig and Serapioni [5] assuming $w(x)$ (the weight) to be in the Muckenhoupt class A_2 [6].

A nonnegative measurable function w in R^n is said to be an A_2 weight if there exists a positive constant k such that(1)

$$⨍_C w(x)\,dx ⨍_C [w(x)]^{-1}\,dx \leq k \qquad (2)$$

for any cube C in R^n.

---

(1) $⨍_C w(x)\,dx = \frac{1}{|C|}\int_C w(x)\,dx$, where $|C|$ stands for the Lebesgue measure in R^n of C.

As shown by the results in [5] the solutions to the equation

$$-\frac{\partial}{\partial x_i}\left(a_{ij}(x)\frac{\partial u}{\partial x_j}\right) = f \tag{3}$$

(where $a_{ij}(x)\xi_i\xi_j$ is equivalent to $w(x)|\xi|^2$, $w(x) \in A_2$) behave essentially in the same way as in the nondegenerate case. In particular a Harnack inequality is true for nonnegative solutions of (3) which in turn implies the local Hölder continuity of solutions.

The situation is quite different in the parabolic case. Indeed, even assuming, in equation (1), $w(x,t)$ to be an A_2 weight in x uniformly in t and the same in t uniformly in x (for precise statements of the assumptions see [2]) one may have unbounded solutions of (1). This fact is in striking contrast with the smoothing behaviour of nondegenerate parabolic operators.

Let us recall a simple example from [2]. Consider in $Q = \{(x,t): |x| < 1, 0 < t < 1\}$ the function $u(x,t) = e^{-a(n-a)t}|x|^{-a}$ $(0 < a < \frac{n}{2})$.

Assume $w(x,t) = w(x) = |x|^2$. $w(x) \in A_2$ for $n \geq 3$. Then

$$\int_Q u^2 w \, dx \, dt + \int_Q |\nabla u|^2 w \, dx \, dt < +\infty$$

and u is a weak solution(2) in Q of

$$\frac{\partial}{\partial x_i}\left(|x|^2 \frac{\partial u}{\partial x_i}\right) = \frac{\partial u}{\partial t}.$$

However, u is unbounded in Q.

This shows that no local regularity result is possible for solutions of equation (1) only under the assumption that $w(x,t)$ is an A_2 weight (in x uniformly in t, etc.).

In [2] it is also shown that the <u>usual</u> Harnack inequality for parabolic operators does not hold for (1).

To make this point clear let us introduce some notation and assume in the following $f \equiv 0$ in equation (1).

Assume to be given in Q a subcylinder

$$Q_\rho(x_o,t_o) = B_\rho(x_o) \times]t_o-\rho^2, t_o+\rho^2[, \ \rho > 0,$$

(2) Precise definitions may be found in [2].

and let

$$Q_\rho^+ = B_{\rho/2}(x_o) \times \,]t_o+\alpha\rho^2, t_o+\rho^2[, \quad Q_\rho^- = B_{\rho/2}(x_o) \times \,]t_o-\beta\rho^2, t_o-\gamma\rho^2[$$

for some fixed numbers $0 < \alpha < 1$, $0 < \gamma < \beta < 1$.

Then, in the nondegenerate case, the following Harnack inequality is true. Assume u is a weak solution of (1) in Q, which is nonnegative in Q_ρ. Then

$$\operatorname*{ess\,sup}_{Q_\rho^-} u(x,t) \leq k \operatorname*{ess\,inf}_{Q_\rho^+} u(x,t) \tag{4}$$

for some constant k independent from ρ and from u.

Some examples in [2] prove that an inequality like (4) may be false for solutions of equation (1) with weights having very "small" degeneracies.

It is however possible to recover a somewhat satisfactory local regularity theory for equation (1) from a slightly different form of the Harnack inequality which is, in some sense, adapted to the degeneracy. Obviously, stronger integrability hypotheses on the weight are needed than A_2 in order to avoid situations like the one mentioned above for the weight $|x|^2$.

From now on assume $w(x,t) = w(x)$. Also assume $w(x)$ satisfies the following $A_{1+\frac{2}{n}}$ condition, similar to (but stronger than) (2)

$$⨍_C w(x)\,dx \left(⨍_C [w(x)]^{-\frac{n}{2}}\,dx\right)^{\frac{2}{n}} \leq k.$$

Define, for any $x_o \in \Omega$, the continuous and increasing function

$$h_{x_o}(\rho) = \left(\int_{B_{x_o}(\rho)} \left[\frac{1}{w(x)}\right]^{\frac{n}{2}} dx\right)^{\frac{2}{n}}. \tag{5}$$

Let $\tilde{Q}_\rho(x_o,t_o) = B_\rho(x_o) \times \,]t_o-h_{x_o}(\rho), t_o+h_{x_o}(\rho)[$ and similar[(3)] definitions for $\tilde{Q}_\rho^-$ and $\tilde{Q}_\rho^+$.

---

(3) As in the Harnack inequality for nondegenerate operators above but substituting $h_{x_o}(\rho)$ for ρ^2 everywhere. For more details see [3].

Then the following inequality

$$\operatorname*{ess\,sup}_{\tilde{Q}_\rho^-} u(x,t) \leq k \operatorname*{ess\,inf}_{\tilde{Q}_\rho^+} u(x,t) \tag{6}$$

is true for any nonnegative solution u of equation (1) with constant k independent from ρ and u.

We wish to point out that $h_{x_o}(\rho)$ is comparable to ρ^2 if the weight does not degenerate near the point x_o

As it was clear after the work [3] had been completed, (6) implies the local Hölder continuity of solutions of equation (1) with an Hölder exponent which obviously depends on the rate of zero of $h_{x_o}(\rho)$.

Before concluding we wish to mention a different way for obtaining some regularity results for degenerate parabolic equations.

Consider in the cylinder Q the equation

$$\frac{\partial}{\partial x_i}\left(a_{ij}(x)\frac{\partial u}{\partial x_j}\right) = \frac{\partial(w(x)u)}{\partial t} \tag{7}$$

where

$$\frac{1}{\lambda} w(x)|\xi|^2 \leq a_{ij}(x)\xi_i\xi_j \leq \lambda w(x)|\xi|^2 \ \forall\xi \in R^n, \quad \text{a.e. in } \Omega$$

for some positive λ and A_2 weight $w(x)$.

It is shown in [1] that the solutions of (7) are locally bounded in Q. Moreover in the work [4] it is proved that the nonnegative solutions of (7) satisfy the usual Harnack inequality (4) exactly as in the nondegenerate case.

The good behaviour of (7) may be seen as a consequence of the fact that the presence of the weight on the right side kills the evolution part as well as the smoothing elliptic part of the operator.

Frasca and Serapioni are now studying equations slightly more general than (7) and (1), in which two different weights $w_1(x)$ on the left and $w_2(x)$ on the right hand side are allowed. The behaviour of such an equation is close to that of equation (1). However, it is a somewhat involved technical problem to express in a reasonably compact form the assumptions on the weights $w_1(x)$ and $w_2(x)$ which are sufficient to obtain a Harnack inequality.

REFERENCES

1. Chiarenza, F. and M. Frasca: Boundedness for the Solutions of a Degenerate Parabolic Equation, Applicable Anal. 17 (1984), 243-261.
2. Chiarenza, F. and R. Serapioni: Degenerate Parabolic Equations and Harnack Inequality, Ann. Mat. Pura Appl. 137 (1984), 139-162
3. Chiarenza, F. and R. Serapioni: A Harnack Inequality for Degenerate Parabolic Equations, Comm. Partial Differential Equations 9 (1984), 719-749.
4. Chiarenza, F. and R. Serapioni: A Remark on a Harnack Inequality for Degenerate Parabolic Equations, Rend. Sem. mat. Univ. Padova 73 (1985), 179-190.
5. Fabes, E., C. Kenig and R. Serapioni: The Local Regularity of Solutions of Degenerate Elliptic Equations, Comm. Partial Differential Equations 7 (1982), 77-116.
6. Muckenhoupt, B.: Weighted Norm Inequality for the Hardy Maximal Function, Trans. Amer. Math. Soc. 165 (1972), 207-226.

F. Chiarenza
Dipartimento di Matematica
Viale A. Doria 6
95125 Catania
Italia

G DA PRATO & A LUNARDI

Periodic solutions for autonomous fully nonlinear equations in Banach spaces

We are concerned with periodic solutions of the problem

$$u'(t) = f(\lambda,u(t)) \tag{1}$$

where $f :]-1,1[\times D \to X$, $(\lambda,x) \to f(\lambda,x)$ is a C^∞ function, D and X are real Banach spaces with D continuously embedded in X, $f(\lambda,0) = 0$ for $\lambda \in]-1,1[$.

Problem (1) has been extensively studied when f is semilinear (see [2], [4], [5], [7]) and in the finite dimensional case (see for instance [1],[6]).

We assume that the operator $A = f_x(0,0) : D \to X$ generates an analytic semigroup e^{tA} in X, and moreover

$$\begin{cases} i \text{ is a simple isolated eigenvalue of } \tilde{A},\ ki \in \rho(\tilde{A}) \\ \text{for } k \in \mathbb{Z},\ k \neq \pm 1 \\ \\ 1 \text{ is a semisimple isolated eigenvalue of } e^{2\pi\tilde{A}}, \\ \text{with algebraic multiplicity 2.} \end{cases} \tag{2}$$

Here $\tilde{A} : \tilde{D} \to \tilde{X}$, $\tilde{A}(x+iy) = Ax + iAy$, denotes the complexification of A, and $\tilde{D},\tilde{X}$ are the usual complexification of D and X respectively.

Under assumptions (2), the linear problem $u'(t) = Au(t)$ has nontrivial 2π-periodic solutions; thus we look for solutions of (1) with period $2\pi\rho$, ρ being close to 1. Changing t with t/ρ, our problem reduces to find 2π-periodic solutions to the equation

$$u'(t) = \rho f(\lambda,u(t)). \tag{3}$$

It is convenient to introduce the following notations.

If $\gamma \in]0,1[$ and B is a Banach space, $C^\gamma_\#(B)$ denotes the space of all γ-Hölder continuous 2π-periodic functions $\varphi : \mathbb{R} \to B$, endowed with the norm

$$\|\varphi\|_{C^\gamma_\#(B)} = \sup_{0\le t\le 2\pi} \|\varphi(t)\|_B + \sup_{0\le s\le t\le 2\pi} (t-s)^{-\gamma}\|\varphi(t)-\varphi(s)\|_B .$$

Moreover, $C^{1,\gamma}_\#(B)$ is the space of the differentiable functions $\varphi : \mathbb{R} \to B$

such that φ and φ' belong to $C^{\gamma}_{\#}(B)$. $C^{1,\gamma}_{\#}(B)$ is endowed with the norm

$$\|\varphi\|_{C^{1,\gamma}_{\#}(B)} = \sup_{0\le t\le 2\pi} \|\varphi(t)\|_B + \|\varphi'\|_{C^{\gamma}_{\#}(B)} .$$

To solve equation (3) we set

$$\begin{cases} F :]-1,1[\times]0,2[\times C^{\gamma}_{\#}(D) \cap C^{1,\gamma}_{\#}(X) \to C^{\gamma}_{\#}(X) \\ F(\gamma,\rho,u)(t) = u'(t) - \rho f(\lambda,u(t)). \end{cases} \tag{4}$$

Then F is of class C^{∞}. We shall study the equation

$$F(\lambda,\rho,u) = 0 \tag{5}$$

by classical implicit function arguments. To this aim we have to characterize the kernel and the range of the derivative $F_u(0,1,0)$.

<u>LEMMA.</u> *Let $x_o, y_o \in D\backslash\{0\}$ be such that $\tilde{A}(x_o \pm iy_o) = \pm i(x_o \pm iy_o)$, and denote by X_o the subspace of X generated by x_o and y_o. Then we have:*

$$\text{Ker } F_u(0,1,0) = \{e^{tA}x; x \in X_o\}; \ \dim \text{Ker } F_u(0,1,0) = 2$$

$$\text{Range } F_u(0,1,0) = \{z \in C^{\gamma}_{\#}(X);\ Q\int_0^{2\pi} e^{(2\pi-s)A}z(s)ds = 0\}$$

where $Q : X \to X_o$ is the projector defined by

$$Qx = \frac{1}{2\pi i}\int_{\Gamma}(\xi-\tilde{A})^{-1}x d\xi, \quad x \in X,$$

Γ being a suitable path around $\{+i,-i\}$.

The proof relies on maximal regularity properties for the solution of the initial value problem

$$\begin{cases} v'(t) = Av(t) + z(t), \quad t \ge 0 \\ v(0) = x \end{cases} \tag{6}$$

and can be found in [3].

In order to state our main result, we need a transversality assumption on the eigenvalues of the operator $A(\lambda) = f_x(\lambda,0)$. It can be shown that there exist $\lambda_o \in]0,1[$ and two functions $\alpha,\beta \in C^{\infty}(]-\lambda_o,\lambda_o[;\mathbb{R})$ such that $\alpha(0) = 0$,

$\beta(0) = 1$ and $\alpha(\lambda) \pm i\beta(\lambda)$ are simple eigenvalues of $\tilde{A}(\lambda)$.

THEOREM. *Under the previous hypotheses, assume moreover that* $\alpha'(0) \neq 0$. *Then there exist* $\sigma_o > 0$ *and three functions* $\lambda :]-\sigma_o,\sigma_o[\to]-1,1[, \rho :]-\sigma_o,\sigma_o[\to]0,2[$, $u :]-\sigma_o,\sigma_o[\to C^{\gamma}_{\#}(D) \cap C^{1,\gamma}_{\#}(X)$, *such that* $\lambda(0) = 0$, $\rho(0) = 1$, $u(0) \equiv 0$, *and* $u(\sigma)$ *is a solution of* (3), *with* $\rho = \rho(\sigma)$, $\lambda = \lambda(\sigma)$. *Moreover* $u(\sigma)$ *is nonconstant for* $\sigma \neq 0$.

The theorem may be proved using the previous Lemma and classical arguments of bifurcation; the proof is in [3], where the uniqueness of $u(\sigma)$ in the class of the small solutions of (3) is also proved.

Let us give now an example.

EXAMPLE. Consider the equation

$$u_t(t,x) = \varphi(\lambda,u(t,x),u_x(t,x),u_{xx}(t,x)) \tag{7}$$

where $\varphi :]-1,1[\times \mathbb{R}^3 \to \mathbb{R}$, $(\lambda,p_1,p_2,p_3) \to \varphi(\lambda,p_1,p_2,p_3)$ is a C^∞ function such that $\varphi(\lambda,0,0,0) = 0$. We assume that equation (7) is parabolic for λ and u small, that is

$$\varphi_{p_3}(0,0,0,0) > 0.$$

Equation (7) can be written in the abstract form (1), with

$$X = C_{\#}(\mathbb{R}), \quad D = C^2_{\#}(\mathbb{R})$$

$$f(\lambda,v) = \varphi(\lambda,v(\cdot),v'(\cdot),v''(\cdot)).$$

The spectrum of the operator $A = f_v(0,0)$ consists of the eigenvalues $\{\varphi_{p_1}(0,0,0,0) + ik\varphi_{p_2}(0,0,0,0) - k^2\varphi_{p_3}(0,0,0,0);\ k \in \mathbb{Z}\}$. If for some $k \in \mathbb{Z}$ we have

$$\varphi_{p_1}(0,0,0,0) = k^2\varphi_{p_3}(0,0,0,0), \quad k\varphi_{p_2}(0,0,0,0) = 1$$

then condition (2) holds. Moreover, if

$$\varphi_{p_1\lambda}(0,0,0,0) \neq k^2\varphi_{p_3\lambda}(0,0,0,0),$$

then the transversality assumption holds. Therefore there exists a branch of nontrivial solutions of (7), which are 2π-periodic with respect to both t and x.

REFERENCES

1. Chow, S.N. and J.K. Hale: Methods of bifurcation theory (Springer, 1982).
2. Crandall, M.G. and P.H. Rabinowitz: The Hopf bifurcation theorem in infinite dimensions, Arch. Rational Mech. Anal. 68 (1978), 53-72.
3. Da Prato, G. and A. Lunardi: Hopf bifurcation for fully nonlinear equations in Banach space, Anal. Non Lin. (to appear).
4. Henry, D.: Geometric theory of semilinear parabolic equations. Lecture Notes in Mathematics 840 (Springer, 1981).
5. Ize, J.: Periodic solutions of nonlinear parabolic equations, Comm. Partial Differential Equations 4 (1979), 1299-1387.
6. Marsden, J. and M.F. McCracken: The Hopf bifurcation and its applications (Springer, 1976).
7. Negrini, P. and A. Tesei: Attractivity and Hopf bifurcation in Banach spaces. J. Math. Anal. Appl. 78 (1980), 204-221.

G. Da Prato
Scuola Normale Superiore
Piazza dei Cavalieri 7
56100 Pisa
Italia

A. Lunardi
Dipartimento di Matematica
Università di Pisa
Via Buonarroti 2
56100 Pisa
Italia

R DAL PASSO & S LUCKHAUS

A degenerate diffusion equation not in divergence form

In this paper we look for nonnegative solutions with compact support of the equation

$$u_t = u\Delta u + g(u) \qquad \text{in} \quad \mathbb{R}^+ \times \mathbb{R}^n. \tag{1}$$

We show that there is a unique maximal solution. The support of this maximal solution coincides for all times with the support of the initial values. On the other hand one can find for arbitrary time t, positive, solutions whose support vanishes at time t. Finally, in most cases the maximal solution is characterized by the property that it remains positive on those points where the initial value is positive. Detailed proofs will appear elsewhere [2].

1. Introduction

Take the equation

$$u_t = c \cdot \nabla u + u\Delta u + g(u) \qquad \text{in} \quad \mathbb{R}^+ \times \mathbb{R}^n \tag{2}$$

$$u(0,\cdot) = u_o \geq 0 \qquad \text{in} \quad \mathbb{R}^n,$$

where u_o has compact support. We define the notion of a weak solution to (2). We construct then weak solutions under suitable assumptions on g, such that the support of u does not grow. Note that (2) is just (1) in a coordinate system moving with constant speed. So we have constructed a solution of (1) whose support drifts away from the original support at arbitrary speed c. This is possible because of the strength of the degeneracy $u\Delta u$, which thus proves to be worse than a degeneracy $\Delta\varphi(u)$. Next we use the fact that (1) can be written in divergence form where u is positive, to derive a comparison result between solutions and positive solutions. This gives the uniqueness of the maximal solutions, and shows that the "drift-solutions" die out in arbitrary time. Finally with the same comparison technique we show that for initial data whose support is nice and who are positive in its interior, the maximal solution is the only one, which remains positive in the interior of the support of u_o. For that statement, the requirement that u_o have no zeros in the interior of $\text{supp}(u_o)$ cannot be dropped. We would like to mention that equation (1) has been studied also by M. Ughi in [3].

2. Statement of the results

First we give a definition of a weak solution for (2). Since we shall need them for existence, we assume throughout the following for u_o and g:

$$0 \leq u_o \in L^\infty, \quad u_o \equiv 0 \quad \text{in } \partial\Omega \tag{3}$$

where Ω is a bounded domain with Lipschitz boundary;

$$|g'(u)| < K|u|, \quad |g(u)| < K|u| + \lambda|u|^2: \tag{4}$$

here K is arbitrary and λ is less than the first eigenvalue of $-\Delta$ in Ω.

<u>DEFINITION.</u> $0 \leq u \in L^\infty \cap L^2_{loc}(\mathbb{R}^+, H_o^{1,2}(\Omega))$ *is said to be a weak solution of* (2), *iff for all* $\varphi \in L^\infty \cap H^{1,2}(\mathbb{R}^+ \times \mathbb{R}^n)$ *with bounded support*

$$\int_{\mathbb{R}^+\times\Omega} (u_o-u)\partial_t\varphi + \int_{\mathbb{R}^+\times\Omega} c\cdot\nabla u\varphi + \int_{\mathbb{R}^+\times\Omega} \nabla u\ \nabla(u\varphi) = \int_{\mathbb{R}^+\times\Omega} g(u)\varphi. \tag{5}$$

Let us state immediately the following extension result.

<u>LEMMA.</u> *If* u *is a weak solution of* (2) *in* Ω *and* $\Omega' \supset \Omega$, *then* u *extended by zero is a solution of* (2) *in* Ω'.

We have the following existence result.

<u>THEOREM 1.</u> *There exists under assumptions* (3) *and* (4) *a weak solution* u *of* (2) *in* Ω' *for all* $\Omega' \supset \Omega$.

Also for every solution of equation (1), i.e. in the case of no drift, we have that the support of u does not grow.

<u>THEOREM 2.</u> *Suppose* u *is a weak solution of* (2), $c = 0$, (3) *and* (4) *hold. Then* $\text{supp}\ u(t,\cdot) \subset \overline{\Omega}$ *for all times* t.

Theorem 2 shows already that for c arbitrarily large the solutions existing according to Theorem 1 have an arbitrarily small extinction time. On the other hand we have a comparison and consequently a uniqueness theorem for a certain type of solution.

THEOREM 3. *Suppose* $u_o \in H_o^{1,2}(\tilde{\Omega})$, $\tilde{\Omega}$ *is such that* $H_o^{1,2}(\tilde{\Omega})=\{u\in H^{1,2}(\mathbb{R}^n)\,|\,u \equiv 0$ *in the interior of* $\tilde{\Omega}\}$. *Suppose* v *is a weak solution of* (1) *such that* $v(0,x) \geq u_o(x)$ *for all* x, $\partial_t v \in L^2$ *and* $v(t,x) \geq \delta(T,\hat{\Omega}) > 0$ *for all* $x \in \hat{\Omega}$, $0 \leq t \leq T$ *where* $T < \infty$, $\hat{\Omega} \subset\subset \tilde{\Omega}$.

Then for any weak solution u *of* (1) *we have* $u \leq v$.

COROLLARY. *Suppose* $u_o \in H^{1,2}$; *then there exists a unique maximal solution of* (1). *If* u_o *has no zeros in the interior of its support* $\tilde{\Omega}$ *and* $\tilde{\Omega}$ *fulfils the condition of Theorem 3, then the maximal solution is uniquely characterized by the property that it also has no zeros in* $\tilde{\Omega}$.

REMARK. Even for one space dimension and smooth initial data, if u_o has zeros in $\tilde{\Omega}$, and the second derivative of u_o is large enough at these points, the maximal solution will become positive in $\tilde{\Omega}$ after finite time. On the other hand one can force u to stay zero at these points. So the characterization fails.

3. Outline of the proofs

There is nothing to prove in the Lemma since we did not require φ to vanish on the boundary of Ω.

In Theorem 1, we construct a solution for (2) as follows. First we solve

$$\partial_t u_\varepsilon + c\nabla u_\varepsilon - u_\varepsilon \nabla u = g(u_\varepsilon) + K\varepsilon^2 \tag{6}$$

with boundary data ε and initial values $u_o + \varepsilon$.

We observe that u_ε is a monotone decreasing sequence, $|u_\varepsilon|_\infty$ and also $\int \frac{|\nabla \; \varepsilon|^2}{u_\varepsilon^\alpha}$ are uniformly bounded if $\alpha < 1$. If u denotes the weak limit of u_ε, using as a test function $u_\varepsilon^2 - u^2 - \varepsilon^2$ we find that $\int|\nabla(u_\varepsilon^2 - u^2)|^2$ converges to zero. Taking all the above together, $u_\varepsilon - \varepsilon$ converges strongly to u in $L_2(0,T\ ;\ H_0^{1,2}(\Omega))$. To derive equation (5) for u, we only have to control $-\int \varphi\, \varepsilon\, \Delta u_\varepsilon = +\varepsilon \int \nabla\varphi\, \nabla u_\varepsilon + \varepsilon \int_{\partial\Omega} |\partial_\nu u_\varepsilon| \varphi$. But $\int_{\partial\Omega} |\partial_\nu u_\varepsilon| \leq \mathrm{const}\left(1 + \ln\left(\frac{1}{\varepsilon}\right)\right)$ which proves the claim.

For Theorem 2 it suffices to take $\frac{1}{u+\delta}\varphi$ as a test function where φ has support in $c\,\overline{\Omega}$, to show that $u\varphi \equiv 0$.

Theorem 3 is a little bit more tricky. It makes use of the following identity:

$$\int_0^t \langle \partial_t(u-v),\ H_\delta(e^u - e^v) \rangle = \int_\Omega \int_0^u H_\delta(e^z - e^v)dz\Big|_0^t - \tag{7}$$

$$- \int_0^t \int_\Omega \partial_t v \int_0^u (e^z - e^v)_+ H'_\delta(e^z - e^v)dz;$$

observe that $\partial_t(u-v) \in (L^\infty \cap H_0^{1,2})^*$ and $\partial_t v \in L^2$.

Here $H_\delta(x) = \sup(0, \inf(1, \frac{x}{\delta}))$.

One divides equation (1) by u (v respectively), takes the difference of the equations and multiplies by $\psi_\varepsilon\, H_\delta(u-v)$, where ψ_ε is an approximation of the first eigenfunction of $-\Delta$ in $\tilde{\Omega}$ with compact support in $\tilde{\Omega}$. That gives the result.

For the Corollary one just has to show that for c = 0 the solution u as constructed in Theorem 1 has the properties required of v in Theorem 3. The important thing to note is that for c = 0, $\frac{\partial_t u_\varepsilon}{u_\varepsilon}$ fulfils the minimum principle.

Details of these calculations and the ones justifying the Remark are to be found in [2].

REFERENCES

1. Brezis, H. and M.G. Crandall: Uniqueness of solutions of the initial value problem for $u_t - \Delta\varphi(u) = 0$, J. Math. Pures Appl. 58 (1979), 153-163.

2. Dal Passo, R. and S. Luckhaus: On a degenerate diffusion problem not in divergence form (preprint SFB 123 310, 1985).

3. Ughi, M.: A degenerate parabolic equation modelling the spread of an epidemic (preprint Università di Firenze, 1984).

R. Dal Passo
Istituto per le Applicazioni del Calcolo "M. Picone"
Consiglio Nazionale delle Ricerche
Viale del Policlinico 137
00161 Roma
Italia

S. Luckhaus
Sonderforschungsbereich 123
Im Neuenheimer Feld 294
6900 Heidelberg
BRD

E DI BENEDETTO

Periodic solutions of the dam problem

1. Introduction

This is an informal account of some results obtained jointly with A.Friedman [1], concerning the existence of periodic solutions for the dam problem. Even though some of our results hold for general dams, we have chosen here, for simplicity of presentation, to restrict ourselves to rectangular dams.

Let $b > 0$ be given and let $D \equiv \{0 < x < 1\} \times \{0 < y < b\}$ be the dam and for $0 < T < \infty$ set $D_T \equiv D\times(0,T]$.

In D_T we seek, formally, to solve the boundary value problem:

$$\frac{\partial}{\partial t}(\alpha v + H(v)) - \Delta v \ni \frac{\partial}{\partial y} H(v) \quad \text{in } \mathcal{D}'(D_T) \tag{1.1}$$

$$v(0,y,t) = (y_1(t) - y)^+; \quad v(1,y,t) = (y_2(t) - y)^+, \; t \in (0,T) \tag{1.2}$$

$$\left(\frac{\partial}{\partial y} v\right)(x,0,t) = -1, \quad x \in (0,1), \quad t \in (0,T) \tag{1.3}$$

$$\alpha v(x,y,0) + H(\alpha v(x,y,0)) \ni \alpha v_0(x,y) + \xi_0(x,y). \tag{1.4}$$

Physically (1.1) - (1.4) models the time dependent flow of a viscous fluid, of hydrostatic potential $v+y$, in a porous dam. Here $t \to y_i(t)$, $t \in (0,T)$ are prescribed levels, (1.3) signifies that the bottom of the dam is impervious and $H(\cdot)$ is the Heaviside graph

$$H(s) \equiv \begin{cases} 1, & s > 0 \\ [0,1], & s = 0 \\ 0, & s < 0 \end{cases}.$$

If $\alpha > 0$ the fluid is compressible and if $\alpha = 0$ the fluid is incompressible. We shall treat side by side the cases $\alpha > 0$ and $\alpha = 0$.

To make the meaning of (1.1) - (1.4) precise, we recall that the stationary dam problem consists of finding a pair (v,ξ) such that:

$$
(SP)\begin{cases}
v \in H'(D),\ \xi \subset H(v) \text{ (in the sense of the graphs);}\\
v \geq 0 \quad \text{in} \quad D;\\
v(0,y) = y_1;\ v(1,y) = y_2;\ 0 < y_i < b,\ i = 1,2 \text{ (in the sense of the traces);}\\
\int_D (\nabla v + \xi\vec{e})\ \nabla\varphi dx\, dy \leq 0,\ \vec{e} \equiv (0,1),\ \forall\varphi \in H^1(D);\\
\varphi \geq 0 \quad \text{on} \quad \{y = 0\},\ \varphi(0,y) = \varphi(1,y) = 0 \text{ (in the sense of the traces).}
\end{cases}
$$

It is well-known that (SP) has a unique solution and if $\varepsilon < y_i < b-\varepsilon$, $i = 1,2$, then

$$
\begin{cases}
v(x,y) > 0,\ \xi(x,y) = 1, & (x,y) \in (0,1) \times (0,\varepsilon);\\
v(x,y) \equiv 0,\ \xi(x,y) = 0, & (x,y) \in (0,1) \times (b-\varepsilon,b).
\end{cases}
$$

On physical grounds the dam is wet in a neighbourhood of the bottom (i.e. $v > 0$, $\xi = 1$ near $y = 0$) and dry in a neighbourhood of the top (i.e. $v = 0$, $\xi = 0$ near $y = b$). We call $\mathcal{C}_\varepsilon$ the class of solutions of (1.1) - (1.4) having such a property, and will seek solutions in $\mathcal{C}_\varepsilon$.

We assign boundary data $t \to y_i(t)$ as Lipschitz continuous functions in $[0,T]$ satisfying

$$0 < \varepsilon \leq y_i(t) \leq b-\varepsilon,\ t \in [0,T],\quad i = 1,2,$$

and initial data

$$\alpha v_0 \in L^\infty(D), \quad \alpha v_o \geq 0,$$

$$\xi_o(x,y) \in H(\alpha v_0) \quad \text{satisfying:}$$

$$\xi_0(x,y) = 1 \quad \text{if} \quad (x,y) \in (0,1) \times (0,\varepsilon),$$
$$\xi_0(x,y) = 0 \quad \text{if} \quad (x,y) \in (0,1) \times (b-\varepsilon,\varepsilon),$$

for some $\varepsilon > 0$. Then we define our concept of weak solution of (1.1) - (1.4) as follows. A weak solution of (1.1) - (1.4) is a pair of measurable functions (v,ξ) such that:

(i) $\xi \subset H(v)$(graphs); $t \to \xi(\cdot,t) \in C(0,T;H^{-1}(D))$;

(ii) $\alpha v \in C(0,T;L^2(D))$; $v \in L^2(0,T;H^1(D))$, $v \geq 0$ in D_T;

(iii) $v > 0$ in $(0,1) \times (0,\varepsilon)$; $v = \xi = 0$ in $(0,1) \times (b-\varepsilon,b)$;

(iv) v satisfies (1.2) in the sense of the traces and (1.3) in the classical sense;

(v) for any $\psi \in H^1(D_T)$ with $\psi(x,y,T) = 0$, $\psi(0,y,t) = \psi(1,y,t) = 0$, $t \in (0,T)$; $\psi(x,0,t) \geq 0$, the following integral inequality holds:

$$\iint_{D_T} \{-(\alpha v+\xi)\psi_t + \nabla v \nabla \psi + \xi \psi_y\} dX dt \leq \int_D (\alpha v_o+\xi_o)\psi(X,0)dX, \tag{1.5}$$

where $X = (x,y)$.

Notice that (1.5) implies that $\alpha v_t - \Delta v = 0$ in $[v > 0]$ and therefore (iii) implies that v is classical in a D_T-neighbourhood of $\mathcal{y} = 0$; consequently, it makes sense to require that (1.3) holds in the classical sense.

2. Existence and uniqueness

It is shown in [1] that solutions in $\mathcal{C}_\varepsilon$ do exist. A feature of our existence theorem consists in comparing at each time $t \in (0,T)$ the solution of the time-dependent problem, with the solutions of suitably constructed stationary problems. We find solutions for which

(i) $\|v\|_{L^\infty(D_T)} \leq M$, where M depends upon the data and $\alpha\|v_o\|_{L^\infty(D)}$, and it is independent of α;

(ii) $\sqrt{\alpha}\, v_t \in L^2_{loc}(D_T)$;

(iii) if $\alpha > 0$, $v \in C(\overline{D} \times \{t \geq \eta\})$, $\forall \eta > 0$, and if $v_o \in C(\overline{D})$ then $v \in C(\overline{D}_T)$;

(iv) let $\alpha \geq 0$. If $v_o \in C^1(D)$ and $\frac{\partial}{\partial y} v_o \leq 0$, then $v_y \leq 0$ in $\mathcal{D}'(D_T)$. In particular the free boundary $\partial[v > 0] \cap D_T$ is a graph in the y-direction;

(v) if (v_o, ξ_o) coincide with the unique solution of a stationary dam problem, and if $\mathcal{y}_i'(t) \geq 0$, $i = 1,2$, then $v_t \geq 0$ in $\mathcal{D}'(D_T)$. In particular, the free boundary is a graph in the t-direction.

An interesting feature of solutions in the class $\mathcal{C}_\varepsilon$ is their uniqueness.

THEOREM 1. *Let $\alpha \geq 0$. Solutions in C_ε are unique.*

Proof. We give a formal proof to convey the main ideas. Suppose (v_j, ξ_j), $j = 1,2$ are two solutions and set

$$w = v_1 - v_2; \qquad H = \alpha w + \xi_1 - \xi_2.$$

Let f be the unique solution of

$$\begin{cases} -\Delta f = H & \text{in } D \text{ for a.e. } t \in [0,T] \\ f = 0 & \text{on } \partial D \cap \{y > 0\} \\ f_y = 0 & \text{on } [0,1] \times [y = 0]. \end{cases} \tag{2.1}$$

Write (1.1) for (v_1, ξ_1), (v_2, ξ_2), take the difference and use (2.1) to obtain

$$\frac{\partial}{\partial t}(-\Delta f) - \Delta w = \frac{\partial}{\partial y}(-\Delta f) - \alpha w_y \quad \text{in } \mathcal{D}'(D_T). \tag{2.2}$$

Multiply formally by f and integrate over D to obtain

$$\frac{1}{2}\frac{\partial}{\partial t}\|\nabla f\|_{2,D}^2(t) + \int_D wH = \int_D f_y \Delta f + \int_D \alpha w\, f_y = \tag{2.3}$$

$$= \frac{1}{2}\int_0^1 \{f_y^2(x,b,t) + f_x^2(x,0,t)\}dx + \alpha\int_D w\, f_y.$$

Since $v_j \in C_\varepsilon$, we have

$$\Delta f = 0 \quad \text{in} \quad (0,1) \times (b-\varepsilon, b). \tag{2.4}$$

Let ζ be a cutoff function in D, $\zeta = \zeta(y)$, $\zeta(b) = 1$, $\zeta(y) = 0$ if $y < b-\varepsilon$, $|\zeta'(y)| < 2/\varepsilon$. Multiplying (2.4) by $f_y\zeta$ and integrating in D we obtain

$$\int_0^1 f_y^2(x,b,t)dx = \int_D \nabla f\, f_y\, \nabla\zeta + \frac{1}{2}\int_D \frac{\partial}{\partial y}|\nabla f|^2\zeta =$$

$$= \int_D \nabla f\, f_y\, \nabla\zeta + \frac{1}{2}\int_0^1 f_y^2(x,b,t) - \frac{1}{2}\int_D |\nabla f|^2\zeta_y.$$

Therefore

$$\int_0^1 f_y^2(x,b,t) \le C\int_D |\nabla f|^2 . \tag{2.5}$$

Now let $\zeta = \zeta(y)$ be a cutoff function in D such that $\zeta(0) = 1$, $\zeta(y) = 0$ if $y > \varepsilon$ and $|\zeta'(y)| \le 2/\varepsilon$, $|\zeta''(y)| \le c/\varepsilon^2$. Since $v_j \in C_\varepsilon$, from (2.1) we have

$$\Delta f = 0 \quad \text{on} \quad \{0 < x < 1\} \times \{0 < y < \varepsilon\}. \tag{2.6}$$

Multiplying by $f\zeta^2$ and integrating over D we get

$$\int_0^1 (f_{yy}f)(x,0,t)dx + \int_D \nabla f_y \nabla f \zeta^2 + \int_D \nabla f_y f \, \nabla\zeta^2 = 0. \tag{2.7}$$

The second integral equals

$$\frac{1}{2}\int_D \frac{\partial}{\partial y} |\nabla f|^2 \zeta^2 = -\frac{1}{2}\int_0^1 f_x^2(x,0,t) - \int_D |\nabla f|^2 \zeta\zeta_y ,$$

and the third integral is equal to

$$\int_D f_{yy} f(\zeta^2)_y = -\int_D f_y^2(\zeta^2)_y - \int_D f_y \, f(\zeta^2)_{yy}.$$

Therefore (2.7) becomes

$$\int_0^1 (f_{yy}f)(x,0,t) - \frac{1}{2}\int_0^1 f_x^2(x,0,t) \le C\int_D |\nabla f|^2. \tag{2.8}$$

But $f_{yy} = -f_{xx}$ on $\{y = 0\}$, so that

$$\int_0^1 f_x^2(x,0,t)dx \le C\int_D |\nabla f|^2 .$$

Substituting these estimates in (2.3) we obtain

$$\frac{\partial}{\partial t} \|\nabla f\|_{2,D}^2(t) + 2\int_D wH \le C\|\nabla f\|_{2,D}^2(t) + \frac{\alpha}{2} \|w\|_{2,D}^2(t). \tag{2.9}$$

Integrate over (0,t) and absorb the last term on the left hand-side to obtain

$$\|\nabla f\|^2_{2,D}(t) \leq C\int_0^t \|\nabla f\|^2_{2,D}(\tau)d\tau.$$

This obviously implies uniqueness. The proof can be made rigorous by a standard time-regularization technique (see [1]).

3. Periodic solutions and large time behaviour

If D is a reservoir, the levels $y_i(t)$ vary according to seasonal variations. Therefore it is of interest to determine whether (1.1) - (1.3) (with $T = \infty$) has a periodic solution.

We assume that the levels $t \to y_i(t)$ are periodic functions in $(0,\infty)$ with period $\sigma > 0$, i.e. $y_i(t+\sigma) = y_i(t)$, for any $t \geq 0$. Moreover we impose

$$\varepsilon < y_i(t) < b - \varepsilon, \quad \text{for any} \quad t \geq 0.$$

Then by various compactness arguments on the Poincaré map we prove that, for $\alpha \geq 0$, (1.1)-(1.3) has at least a periodic solution (v,ξ) in $\mathcal{C}_\varepsilon$ with period σ. That is we find an initial pair $(\alpha v_0,\xi_0)$ such that (i) - (v) of Section 1 hold for any $T > 0$ and moreover

$$\begin{aligned} &v \in C(\overline{D}_\infty); \quad \xi(\cdot,t) \in C(0,\infty;H^{-1}(D)), \\ &v(\cdot,t+\sigma) = v(\cdot,\sigma); \; \xi(\cdot,t+\sigma) = \xi(\cdot,\sigma) \quad \text{for any} \quad t \geq 0, \\ &\qquad\qquad\qquad\qquad \xi(\cdot,\sigma) \subset H(\alpha v(\cdot,\sigma)). \end{aligned}$$

The "periodic" behaviour for large times is a consequence of the following estimate, which holds for any pair of solutions.

Let $t \to y_i(t)$, $\hat{y}_i(t)$, $i = 1,2$, $t \in (0,\infty)$ be assigned levels satisfying:

$$\varepsilon < y_i(t),\ \hat{y}_i(t) < b - \varepsilon,\ t \in (0,\infty),\ i = 1,2.$$

Let $(\alpha v_0,\xi_0)$, $(\alpha\hat{v}_0,\hat{\xi}_0)$ be initial data satisfying $\xi_0 \in H(\alpha v_0)$, $\hat{\xi}_0 \subset H(\alpha\hat{v}_0)$ and

$$v_0 > 0, \quad \hat{v}_o > 0, \quad \xi_o = \hat{\xi}_o = 1 \quad \text{on} \quad (0,1) \times (0,\varepsilon),$$

$$v_0 = \hat{v}_0 = \xi_0 = \hat{\xi}_0 \equiv 0 \qquad \text{on} \quad (0,1) \times (b-\varepsilon,b).$$

Fix T > 0 and let (v,), $(\hat{v},\hat{\xi})$ be the corresponding solutions of (1.1) - (1.4) in D_T. Then for $\alpha > 0$ or $\alpha = 0$ we have

$$\int_0^T \int_D \{|v-\hat{v}|^2 + (\xi-\hat{\xi})(v-\hat{v})\} \quad \leq \tag{3.1}$$

$$\leq C \sum_{i=1}^{2} \int_0^T |y_i(t) - \hat{y}_i(t)| \, dt + C\int_D |\alpha(v-\hat{v}_o) + \xi_o - \hat{\xi}_o|,$$

where C is independent of T.

REMARK. The proof of (3.1) uses in an essential way the fact that our solutions are in C_ε.

We employ (3.1) to study the periodic behaviour of solutions.

Take periodic boundary data $\hat{y}_i(t)$ and denote by $(\hat{v},\hat{\xi})$ a periodic solution (for $\alpha \geq 0$) with

$$\alpha\hat{v}(x,0) = \alpha\hat{v}_o, \quad \hat{\xi}(\cdot,0) = \hat{\xi}_o,$$

constructed previously. Now let $y_i(t)$, αv_o, ξ_o be boundary and initial data with

$$\int_0^\infty |y_i(t) - \hat{y}_i(t)| \, dt < \infty, \tag{3.2}$$

and denote by (v,ξ) the unique solution in D_∞ corresponding to these data. then from (3.1) we deduce that

$$\int_0^\infty \int_D |v-\hat{v}|^2 < \infty;$$

therefore there exists a sequence $\{t_n\}^\infty$ such that

$$\int_D |v(x,t_n)-\hat{v}(x,t_n)|^2 dx \to 0 \quad \text{as} \quad n \to \infty.$$

Moreover if $\alpha > 0$, then

$$\lim_{t\to\infty} \|v(x,t) - \hat{v}(x,t)\|_{\infty,D} = 0.$$

Another consequence of (3.1) is that periodic solutions are unique. Indeed if (v,ξ), $(\hat{v},\hat{\xi})$ are two periodic solutions (with some period σ) corresponding to the same periodic levels $y_i(t)$, then (3.1) implies

$$n \iint_{D_\sigma} |v-\hat{v}|^2 \leq C, \quad \text{for any} \quad n \in N$$

and hence $v = \hat{v}$, $\xi = \hat{\xi}$.

REFERENCES

1. Di Benedetto, E. and A. Friedman: Periodic behaviour for the evolutionary dam problem and related free boundary problems (to appear).

E. Di Benedetto
Department of Mathematics
Northwestern University
Evanston, IL 60201
U.S.A.

J I DÍAZ & J HERNÁNDEZ

Qualitative properties of free boundaries for some nonlinear degenerate parabolic equations

1. Introduction

This paper is a short survey of some recent work by the authors, concerning qualitative properties of nonlinear degenerate parabolic equations. The associated stationary problem was considered by the authors in [7] by using a local comparison technique involving some kind of local radial super-solutions, which was previously introduced by the first author in [5]. There the main interest was the study of the dead core, namely the subset where the (positive) solutions vanish identically; some necessary and/or sufficient conditions for the existence of a (non-empty) dead core, together with additional information about its size and location, were obtained (see [1] and [11] for related work as well as the monograph [6]).

Here we apply the same kind of arguments to a rather large class of nonlinear (possibly) degenerate parabolic equations complemented with non-zero Dirichlet boundary conditions (see Problem (P) below). Some results for the case of pure powers, i.e., $\varphi(u) = u^m$ and $f(u) = u^p$ were obtained in [8]. Here we extend this investigation to nonlinearities φ and f which are not necessarily powers but have only a similar qualitative behaviour (see assumptions (H_1) and (H_2) below) near the origin. We refer the reader to [2] - [4] and [13] - [15] for other related work.

Very roughly speaking, a large part of our results seem to be new in this more general situation, and some of them extend to the case $0 < p < 1$ theorems known for $p \geq 1$. More detailed information can be found below (see also [8][9]). An extended version of this survey, including also work in [8], with full proofs and many complementary results and applications will appear in [9]: in particular, we will give there applications to some reaction-diffusion systems arising in combustion theory (see [2][8]) and population dynamics with nonlinear diffusion ([12]).

2. Main theorems

In this section we consider the following degenerate parabolic problem:

$$\begin{cases} u_t - \Delta\varphi(u) + f(u) = 0 & \text{in } Q = \Omega\times(0,\infty) \\ u(x,t) = h(x,t) & \text{on } \Sigma = \partial\Omega\times(0,\infty) \\ u(x,0) = u_o(x) & \text{in } \Omega, \end{cases} \qquad (P)$$

where Ω is a bounded domain in $\mathbb{R}^N$ with smooth boundary $\partial\Omega$, under the following assumptions:

φ is a continuous increasing function, $\varphi(0) = 0$ and (2.1)
$\varphi' > 0$, $\varphi'' > 0$ in $(0,\infty)$;

f is continuous, $f(0) = 0$; there exists a continuous increasing (2.2)
function f_o such that $0 \leq f_o(s) \leq f(s)$ for every $s \geq 0$;

$h \in L^\infty(\Sigma)$, $h \geq 0$ in Σ; $u_o \in L^\infty(\Omega)$, $u_o \geq 0$ on Ω. (2.3)

Our main result in this section is the following theorem.

THEOREM 2.1. *Suppose that* $u \in C(\overline{Q})$, $u \geq 0$, *is a solution of problem* (P) *with* (2.1) - (2.3). *Moreover assume that*

$$\int_0^1 \frac{ds}{\left[\int_0^s f_o(\varphi^{-1}(t))dt\right]^{1/2}} < +\infty \qquad (H_1)$$

and

$$\int_0^1 \frac{ds}{f_o(s)} < +\infty \qquad (H_2)$$

are satisfied. Then there exists $T_o > 0$ *such that for every* $t \geq T_o$ *we have*

$$N(u(.,t)) \equiv \{x \in \Omega | u(x,t) = 0\} \supset \{x \in \Omega | d(x, \bigcup_{\tau\geq 0} S(h(.,\tau)) \geq L\}$$

where S *denotes the support of the corresponding function, and L is a constant depending on* φ, f_o, h, u_o, Ω *and* N.

The main tool for the proof of Theorem 2.1 is the following Lemma, which generalizes Lemma 2.1 in [7]. Its proof can be found in [6].

LEMMA 2.1. *If we define* $\eta(s) = \psi^{-1} 1/N(s)$, *where*

$$\psi_\mu(r) = \int_0^r \frac{ds}{\left[\mu\int_0^s \frac{1}{2} f_o(\varphi^{-1}(t))dt\right]^{1/2}},$$

then for any $x_o \in \Omega$ *we have*

$$-\Delta\eta(|x-x_o|) + \frac{1}{2} f_o(\varphi^{-1}(\eta(|x-x_o|))) \geq 0 \quad in \quad \Omega. \tag{2.4}$$

Moreover $\eta(0) = \eta'(0) = 0$ *and* $\eta(s) > 0$ *if* $s \neq 0$.

<u>Sketch of the proof of Theorem 2.1.</u> We define (this is an idea adapted from [10])

$$\overline{u}(x,t) = \varphi^{-1}(\eta(|x-x_o|) + \varphi(U(t)))$$

where $\eta(s)$ and $\psi_\mu(r)$ are as in Lemma 2.1 (we remark that by (H_1) we have $\psi_\mu(r) < +\infty$) and U is a positive solution of the ordinary differential equation

$$\frac{dV}{dt} + \frac{1}{2} f_o(V) = 0 \tag{2.5}$$

$$V(0) = \|u_o\|_{L^\infty}.$$

It is not difficult to see that, as a consequence of (H_2), we have $U(t) = 0$ for any $t \geq T_o = \int_0^{\|u_o\|_{L^\infty}} \frac{ds}{2f_o(s)}$.

From (2.1), (2.2) we obtain:

$$\overline{u}_t - \Delta\varphi(\overline{u}) + f_o(\overline{u})$$

$$= \frac{d}{dt}(\varphi^{-1}(\eta(|x-x_o|) + \varphi(U(t))) - \Delta\eta(|x-x_o|) +$$

$$+ f_o(\varphi^{-1}(\eta(|x-x_o|) + \varphi(U(t))) \geq$$

$$\geq \frac{\varphi'(U)}{\varphi'(\varphi^{-1}(\eta+\varphi(U)))}\frac{dU}{dt} - \Delta\eta + \frac{1}{2} f_o(\varphi^{-1}(\eta)) + \frac{1}{2} f_o(U) \geq$$

$$\geq \frac{dU}{dt} - \Delta\eta + \frac{1}{2} f_o(\varphi^{-1}(\eta)) + \frac{1}{2} f_o(U) \geq 0$$

by (2.4) and (2.5), taking into account that

$$\eta+\varphi(U) \geq \varphi(U)$$

implies the inequality

$$\varphi^{-1}(\eta+\varphi(U)) \geq U,$$

hence

$$\varphi'(\varphi^{-1}(\eta+\varphi(U))) \geq \varphi'(U),$$

once again by (2.1).

Concerning the boundary condition, it is easy to show that if we have

$$0 \leq h(x,t) \leq \|h\|_{L^\infty} \leq \varphi^{-1}(\eta(|x-x_0|)) \leq \overline{u}(x,t),$$

then the inequality $h(x,t) \leq \overline{u}(x,t)$ holds at the boundary. Indeed, if $x \notin S(h(.,\tau))$, $h(x,\tau) = 0$ and the inequality is automatically satisfied. If not, it is sufficient that

$$\varphi(\|h\|_{L^\infty}) \leq \eta(|x-x_0|) \text{ for any } x \in \partial\Omega;$$

this is equivalent to

$$\psi_{1/N}[\varphi(\|h\|_{L^\infty})] \leq |x-x_0|$$

or, otherwise stated, be

$$d(x_0, \bigcup_{\tau\geq 0} S(h(.,\tau)) \geq L,$$

where $L = \psi_{1/N}(\varphi(\|h\|_{L^\infty}))$.

As for the initial condition, it is easily seen that

$$0 \leq u_0(x) \leq \|u_0\|_{L^\infty} \leq \varphi^{-1}(\eta(|x-x_0|)) + \varphi(\|a_0\|_{L^\infty})).$$

Thus we obtain (recall (2.2)):

$$\begin{cases} u_t - \Delta\varphi(u) + f_o(u) \leq 0 \leq \overline{u}_t - \Delta\varphi(\overline{u}) + f_o(\overline{u}) & \text{in } Q \\ u(x,t) \leq \overline{u}(x,t) & \text{on } \Sigma \\ u_o(x) \leq \overline{u}(x,0) & \text{in } \Omega ; \end{cases}$$

it follows from comparison results for problem (P) with f_o that $0 \leq u(x,t) \leq \overline{u}(x,t)$. The proof ends by recalling that $u(x_o,t) = 0$ if $t \geq T_o$ and x_o satisfies the above inequality.

REMARK 2.1. It is also possible to prove similar results when replacing $f(u)$ by $c(x,t).f(u)$, with $c(x,t) \geq 0$ (see [8][9]). This seems to be particularly interesting for applications to reaction-diffusion systems.

REMARK 2.2. If $\varphi(s) = s^m$, $f_o(s) = s^p$, then (H_1) is equivalent to $p < m$ and (H_2) is equivalent to $p < 1$. Now, for $m = 1$, (H_1) and (H_2) coincide. But if $\varphi(s) = s$ and f_o is not a power, then (H_1) implies (H_2) but the converse is not true (see [10]).

REMARK 2.3. Our theorem extends some work by Kersner [14] for the case $N = 1$, and also, for $h \equiv 0$ and $\Omega = \mathbb{R}$, results by Kalashnikov [13] and Véron [15] concerning extinction of solutions in finite time. On the other hand, for $m = 1$, $h \equiv 1$, $u_o \equiv 1$, estimates for the dead core as $N(u(.,t)) \supset \{x \in \Omega | d(x,\partial\Omega) \geq L\}$ can be found in [2] (see also [8]).

REMARK 2.4. If (H_2) is satisfied but (H_1) does not hold, it is still possible to get estimates of the kind

$$0 \leq u(x,t) \leq U(t)$$

extending in this way some work by Berstch, Nanbu and Peletier [4], respectively Véron [15]. Similar arguments also allow us to prove the estimate

$$N(u(.,t)) \supset \{x \in \Omega - S(u_o) | d(x, S(u_o) \cup (\bigcup_{\tau \geq 0} S(h(.,\tau))) \geq L'\}$$

for some constant L'.

REMARK 2.5. The same technique of proof allows us also to obtain local (namely depending on the point $x_0 \in \Omega$ and on the norm $\|u_0\|_{L^\infty(B(x_0,\varepsilon))}$) estimates for the extinction time T_0 ($\varepsilon > 0$; see [2],[8],[9]).

THEOREM 2.2. *Assume that* $u \in C(\overline{Q})$, $u \geq 0$, *is a solution of problem* (P) *with* (2.1) - (2.3) *and* (H_1). *If* $x_0 \in \Omega$ *satisfies*

$$0 \leq u_0(x) \leq \varphi^{-1}(\eta(|x-x_0|), 1/N) \tag{2.5}$$

for any $x \in B(x_0,\varepsilon)$, *where* $\varepsilon = \psi_{1/N}(\varphi(M))$, $M = \|u\|_{L^\infty(Q)}$, $\eta(r,\mu) = \psi_\mu^{-1}(r)$, ($\psi_\mu(r)$ *as above*), *then* $u(x_0,t) = 0$ *for any* $t > 0$.

Sketch of the proof. On the set $B(x_0,\varepsilon) \times (0,\infty)$ define the function

$$\overline{u}(x) = \varphi^{-1}(\eta(|x-x_0|), 1/N).$$

Now, reasoning as in [6] we obtain

$$\begin{cases} u_t - \Delta\varphi(u) + f_0(u) \leq 0 \leq \overline{u}_t - \Delta\varphi(\overline{u}) + f_0(\overline{u}) & \text{in } B(x_o,\varepsilon) \times (0,\infty) \\ u(x,0) = u_0(x) \leq \overline{u}(x) = \varphi^{-1}(\eta(|x-x_0|)) & \text{in } B(x_o,\varepsilon) \\ u(x,t) \leq M \leq \overline{u}(x) & \text{on } \partial B(x_0,\varepsilon) \times (0,\infty), \end{cases}$$

where $\|u\|_{L^\infty(Q)} \leq M$. Then a comparison argument gives $0 \leq u(x,t) \leq \overline{u}(x)$.

REMARK 2.6. Theorem 2.2 improves on some results in [4] for $h \equiv 0$; indeed, we only need the local estimate (2.5). If $\varphi(s) = s^m$, $f_0(s) = \lambda s^p$, then

$$u(x) = K_\lambda |x-x_0| \frac{2}{1-f_m}$$

for some $K_\lambda > 0$.

THEOREM 2.3. *Assume* $u \in C(\overline{Q})$, $u \geq 0$, *is a solution of the problem*

$$u_t - \Delta u + f_o(u) = 0 \quad \text{in } Q$$
$$u = 0 \quad \text{on } \Sigma$$
$$u(x,0) = u_o(X) \quad \text{on } \Omega,$$

where (2.2),(2.3) *and* (H_1) *are satisfied. If, moreover,* $u_t \in L^\infty(Q)$, *then we have the estimate*

$$S(u(.,t)) \subset S(u_o) + B(0, \psi_{1/N}(Ct))$$

for any $t > 0$ *and some* $C > 0$, *where* C *depends on* $\|u_t\|_{L^\infty(Q)}$.

Sketch of the proof. Let $t_o > 0$ and $x_o \in S(u(.,t_o)) - S(u_o)$. We consider the region

$$R(t_o) = \{(x,t) | 0 < t < t_o,\ u(x,t) > 0,\ x \notin S(u_o)\}$$

and the function

$$\overline{u}(x) = \eta(|x-x_o|, 1/N).$$

The function $z(x,t) = u(x,t) - \overline{u}(x)$ satisfies

$$z_t - \Delta z + B(x,t)z \leq 0 \quad \text{on} \quad Q$$

for a suitable $B(x,t)$; then the Strong Maximum Principle implies that z takes its maximum on the parabolic boundary of $R(t_o)$. But, on the other hand, $0 = u(x,t) \leq \overline{u}(x)$ for $(x,t) \in \partial_P R(t_o) - S(u_o)$, and $z(x_o,t_o) > 0$. Hence there exists a point $(\overline{x},\overline{t})$ in $\partial S(u_o) \times (0,t_o)$ satisfying $\overline{u(\overline{x})} < \overline{u(\overline{x},\overline{t})}$. This in turn implies

$$d(x_o,S(u(.,t)) \leq |x-x_o| \leq \psi_{1/N}(u(x,t)) \leq \psi_{1/N}(u(\overline{x},\overline{t}) - u(\overline{x},0)) \leq$$

$$\leq \psi_{1/N}(Ct) \leq \psi_{1/N}(Ct_o),$$

which gives the result.

REMARK 2.7. The proof follows an idea of Evans and Knerr [10]. If $\Delta u_0 \in L^\infty(\Omega)$, $u_0 \in H_o^1(\Omega)$ and $h \in L^\infty(\Sigma) \cap H^1(\Sigma)$, then, following a theorem by Bénilan-Ha, $u_t \in L^\infty(Q)$.

REMARK 2.8. If $f_0(s) = s^p$, $0 < p < 1$, then $\psi_{1/N}(Ct) = Ct^{1-p/2}$.

REFERENCES

1. Bandle, C., R. Sperb and I. Stakgold: Diffusion-reaction with monotone kinetics, Nonlinear Anal. TMA 8 (1984), 321-333.

2. Bandle, C. and I. Stakgold: The formation of the dead core in parabolic reaction-diffusion problems, Trans. Amer. Math. Soc. 286 (1984), 275-293.

3. Bertsch, M., R. Kersner and L.A. Peletier: Positivity versus localization in degenerate diffusion equations, Nonlinear Anal. 9 (1985), 987-1008.

4. Bertsch, M., T. Nanbu and L.A. Peletier: Decay of solutions of a degenerate nonlinear diffusion equation, Nonlinear Anal. TMA 8 (1984), 1311-1336.

5. Diaz, J.I.: Técnica de supersoluciones locales para problemas estacionarios no lineales, Memoria no 14 de la Real Academia de Ciencias, Madrid (1980).

6. Diaz, J.I.: Nonlinear Partial Differential Equations and Free Boundaries I:Elliptic equations (Pitman, to appear).

7. Diaz, J.I. and J. Hernández: On the existence of a free boundary for a class of reaction-diffusion systems, SIAM J. Math. Anal. 15 (1984), 670-685.

8. Diaz, J.I. and J. Hernández: Some results on the existence of free boundaries for parabolic reaction-diffusion systems. In "Trends in Theory and Practice of Nonlinear Differential Equations", V.Lakshmikantham Ed., pp.149-156 (Dekker, 1984).

9. Diaz, J.I. and J. Hernández: On the Existence and Evolution of Free Boundaries for Parabolic Reaction-Diffusion Systems (to appear).

10. Evans, L.C. and B.F. Knerr: Instantaneous shrinking of the support of nonnegative solutions to certain nonlinear parabolic equations and variational inequalities, Illinois J. Math. 23 (1979), 153-166.

11. Friedman, A. and D. Phillips: The free boundary of a semilinear elliptic equation. Trans. Amer. Math. Soc. 282 (1984), 153-182.

12. Hernández, J.: Some free boundary problems for reaction-diffusion systems with nonlinear diffusion (to appear).

13. Kalashnikov, A.S.: The propagation of disturbances in problems of nonlinear heat conduction with absorption, Z. Vycisl. Mat. i Met. Fiz. 14 (1974), 891-905.

14. Kersner, R.: The behaviour of temperature fronts in media with nonlinear thermal conductivity under absorption, Vestnik Moskov Univ. Ser. I. Mat. Meh. 33 (1978), 44-51.

15. Véron, L.: Coercivité et proprietés régularisantes des semigroupes non-linéaires dans les espaces de Banach, Publications Mathématiques de Besançon (1977).

J.I. Diaz
Departamento de Ecuaciones Funcionales
Facultad de Matemáticas
Universidad Complutense
28040 Madrid
Espana

J. Hernández
Departamento de Matemáticas
Universidad Autónoma
28049 Madrid
Espana

J I DÍAZ & R KERSNER

The one dimensional porous media equation with convection

1. Introduction

This work* deals with the one dimensional porous media equation with convection. We shall concentrate our attention on nonnegative solutions of the Cauchy problem associated with the simple equation

$$u_t - (u^m)_{xx} - b.(u^\lambda)_x = 0 \qquad (t > 0,\ x \in R), \tag{1}$$

where $m \geq 1$ and $b, \lambda > 0$. Equation (1) (sometimes called the nonlinear Fokker-Planck equation) arises for example in the study of the flow of a fluid through a porous medium. Very roughly speaking, equation (1) describes a fluid moving in a vertical column (in the case of a horizontal column the gravity action is negligible and there is no convection term : $b \equiv 0$). It turns out that the value of the parameter λ is of a great relevance : $\lambda \geq 1$ occurs in downward infiltration problems, $0 < \lambda < 1$ in evaporation type problems (concerning the physical derivation of the equation, we refer the reader to [10], [24]).

From a mathematical point of view, we note that (1) is a quasilinear equation which is nonuniformly parabolic (it is degenerate near the set where $u = 0$) if $m > 1$; moreover, the convection term becomes singular (again where $u = 0$) if $0 < \lambda < 1$. As we shall indicate later, there is an extensive literature concerning the filtration problem ($\lambda > 1$), in contrast with the limited treatment given to the general case ($\lambda > 0$). Our treatment will be general, including the case $0 < \lambda < 1$ (if $\lambda = 1$, equation (1) reduces to the standard porous media equation by an easy change of variables).

In Section 2 we review some results on the existence, regularity and uniqueness of solutions of (1). Section 3 deals with the study of the existence and qualitative behaviour of the free boundaries. Finally, in Section 4 we explain how the previous results can be suitably applied to certain (first order) conservation laws equations, which are hyperbolic, yet have an unbounded dependence domain.

---

*Partially supported by the CAIYT project no. 3383/83.

2. Existence, regularity and uniqueness

Since the equation (1) becomes degenerate or singular near the region $\{u = 0\}$, we cannot expect it to have classical solutions. Several weaker notions of solution may be introduced. We recall the one given in [18] for the Cauchy problem:

(CP) "to find u satisfying (1), as well as $u(0,.) = u_o(.)$",

where $u_o \geq 0$ is a given bounded continuous function on $(-\infty,+\infty)$.

DEFINITION. *A function* u *is a generalized solution of* (CP) *if*

(i) u *is continuous, bounded and nonnegative in* $\overline{Q}$, *where* Q *denotes the strip* $(-\infty,+\infty) \times (0,T]$ *for some fixed* $T > 0$;

(ii) $u(0,x) = u_o(x)$ for any $x \in R$;

(iii) *for every rectangle* $P = [x_o,x_1] \times [t_o,t_1] \subset \overline{Q}$ *and* $\Psi \in C^{2,1}_{x,t}(P)$ *such that* $\Psi(x_1,t) = \Psi(x_2,t) = 0$ for any $t \in [t_o,t_1]$ *we have*

$$0 = I(u,\Psi,P) = \iint_P \{u\Psi_t + u^m\Psi_{xx} - bu^\lambda\Psi_x dxdt - \int_{x_o}^{x_1} u\Psi\Big|_{t_o}^{t_1} dx -$$

$$- \int_{t_o}^{t_1} u^m\Psi_x\Big|_{x_o}^{x_1} dt.$$

The existence of generalized solutions of (CP) may be established by following the constructive method in [21]. Thus we shall obtain a generalized solution as the pointwise limit of a decreasing sequence of classical solutions of (1). This is made in two steps: first, we construct the required sequence; secondly, we study the continuity of the corresponding limit function.

Concerning the first step, a possible choice is the following: $u_k(t,x)$ are classical solutions of (1) in $Q_k = (-k-1, k+1) \times (0,T]$, which satisfy the boundary and initial conditions

$$u_k(\pm(k+1),t) = M = \|u_o\|_\infty \quad \text{and} \quad u_k(x,0) = u_{o,k}(x);$$

here the sequence $u_{o,k}$ tends uniformly to u_o on compact subsets of $(-\infty,\infty)$ and satisfies

$$1/k \leq u_{o,k} \leq M, \quad u_{o,k+1} \leq u_{o,k} \text{ for all } k \geq 1.$$

It follows from a straightforward application of the maximum principle that

$$1/k < u_k \leq M \text{ and } u_{k+1} \leq u_k \quad \text{in } \overline{Q}_k.$$

Then there exists a function u defined on $\overline{Q}$ by $u(x,t) = \lim_{k\to\infty} u_k(x,t)$ for every $(x,t) \in \overline{Q}$. It is easy to see that this function u satisfies condition (iii).

Continuity and other regularity properties of u are proven by obtaining estimates on the modulus of continuity of u_k. One method to get that is to use the Bernstein's method (see the adaptation made in [1]). A crucial point here is choosing some auxiliary functions in a convenient way. In particular, the following result is proven in [10].

THEOREM 1. *Let* $u_o \in C_b(R)$, $u_o \geq 0$, *such that* u_o^β *is Lipschitz continuous for some* $\beta > 0$ *with* $\max\{(m-1),(m-\lambda)^+\} \leq \beta$. *Then there exists at least one generalized solution* u *of* (CP); u *satisfies* $(u^\alpha)_x \in L^\infty(\overline{Q})$, *where* $\alpha=\max\{1,\beta\}$.

REMARK 1. The above result was previously established in [18] and [17] for $\lambda > 1$. We point out that the modulus of continuity of u given in Theorem 1 is optimal ([10]) and that, due to Nash's theorem, $u \in C^\infty$ where $u > 0$. We also note that $(u^m)_x \in L^\infty(Q)$ and u satisfies (CP) in a stronger sense, namely:

$$\iint[\{(u^m)_x - bu^\lambda\}\vartheta_x - u\vartheta_t]dxdt = \int_{-\infty}^{\infty} \vartheta(x,0)u_o(x)dx \tag{2}$$

for any $\vartheta \in C^1(\overline{Q})$ such that $\vartheta(.,T) = 0$, $\vartheta(x,t) = 0$ for $t > 0$ and $|x|$ large enough. Finally, we point out that the regularity assumptions on u_o may be considerably weakened (see [4]).

The uniqueness of generalized solutions, as well as their continuous dependence on the initial data, is a consequence of the following comparison result.

THEOREM 2. *Let* u *be the limit solution constructed in the proof of* Theorem 1. *Let* $\overline{u}$ *(respectively* $\underline{u}$*) be a generalized supersolution (respectively subsolution) of* (CP) *[i.e. satisfying* (i) *and* $I(\overline{u},\zeta,P) \leq 0$ *(respectively* $I(\underline{u},\zeta,P) \geq 0$ *when* $\Psi \geq 0$ *in* (iii)*]. Then for every* $0 < t \leq T$ *we have*

$$\int_{-\infty}^{\infty} (u(x,t) - \overline{u}(x,t))^+ dx \leq \int_{-\infty}^{\infty} (u(x,0) - \overline{u}(x,0))^+ dx \qquad (3)$$

or, respectively,

$$\int_{-\infty}^{\infty} (\underline{u}(x,t) - u(x,t))^+ dx \leq \int_{-\infty}^{\infty} (\underline{u}(x,0) - u(x,0))^+ dx. \qquad (4)$$

In particular, under the assumptions of Theorem 1 there exists a unique generalized solution of (CP).

Inequality (3) (or (4)) is proven previously for each classical solution u_k instead of u; the result follows easily by an already classical duality argument. By approaching suitably the solutions, some simplifications are made without loss of generality. The crucial point of this method is to obtain sharp "a priori" estimates on the test functions $\Psi(x,t)$, which solve a retrograde linear parabolic boundary value problem (see details in [10]).

REMARK 2. Theorem 2 improves on previous uniqueness results, where different restrictions on λ were made (see [18], [17] and [22]). The uniqueness of generalized solutions has been recently investigated in [4] by a different method. The main content of Theorem 2 is a comparison principle: indeed, from (3) it is obvious that $u_0 \leq \overline{u}_0$ implies $u \leq \overline{u}$ on $\overline{Q}$. Finally, we point out that (3) shows that the semigroup associated with the equation (1) is a semigroup of contractions on the space $L^1(\mathbb{R})$ (see [3], [4], [23] and [24] for a different approach).

REMARK 3. The above results on existence, regularity and uniqueness of generalized solutions are, in fact, particular statements of some more general results dealing with the equation

$$u_t = \varphi(u)_{xx} + b(x,u)_x - c(x,u) \qquad (5)$$

under suitable assumptions on φ,b and c. Similar results are also available for other initial-boundary value problems (see [10]). They also can be established for the case of higher space dimension (for an adaptation of the uniqueness argument see [5], [6]).

3. On the free boundaries

Comparison of solutions and conservation of mass, namely

$$\int_{-\infty}^{\infty} u(x,t)dx = \int_{-\infty}^{\infty} u_o(x)dx \qquad (t > 0)$$

([15],[11]) are useful tools in order to study the existence or nonexistence of the free boundaries $\zeta_i(t)$ $(i = 1,2)$, defined by

$$\zeta_1(t) = \inf\{x\in(-\infty,\infty) : u(x,t) > 0\}, \; \zeta_2(t) = \sup\{x\in(-\infty,\infty) : u(x,t) > 0\}.$$

Here we assume that u_o satisfy the conditions of Theorem 1 and moreover

$$\text{supp } u_o = [a_1,a_2] \text{ with } -\infty < a_1 < a_2 < +\infty.$$

The behaviour of the free boundaries $\zeta_i(t)$ is quite different as to whether $\lambda \geq 1$ or $0 < \lambda < 1$.

The case $\lambda > 1$ corresponds to filtration problems and has been widely treated in the literature (see [18], [15] and [16]). In that case both curves $\zeta_i(t)$ are continuous functions on t, whose behaviour may be illustrated by the following properties:

(a) $\zeta_1(t)\downarrow -\infty$ when $t\to\infty$ if $\lambda \geq m$ and $\zeta_1(t)\downarrow a_1 - K$ when $t\to\infty$ if $1 < \lambda < m$, for some $K > 0$;

(b) $\zeta_2(t)\uparrow +\infty$ when $t\to\infty$;

(c) if $u(x_o,t_o) > 0$ for some $(x_o,t_o) \in \overline{Q}$, then $u(x_o,t) > 0$ for every $t \geq t_o$.

When $\lambda = 1$ an adaptation of the Barenblatt-Pattle solution shows that now the properties (a) and (b) are not satisfied.

The case $0 < \lambda < 1$ was considered in [11]. Now there is a strong singularity in the convection term preventing the formation of the free boundary $\zeta_2(t)$, as the following theorem shows.

<u>THEOREM 3.</u> *Let* $0 < \lambda < 1$ *and* u_o *as in Theorem 1. Then the function* $v = u^{m-\lambda}$ *satisfies*

$$v_x > -\frac{Cv}{t} \quad \text{in } Q, \tag{6}$$

where C *is a positive constant only depending on* $m-\lambda, \|u_0\|_\infty$ *and* T. *In particular,* $\zeta_2(t) = +\infty$ *for every* $t > 0$.

The proof of (6) consists in the study of the parabolic equation satisfied by v_x and the application of the comparison principle to a suitable subsolution. (An additional argument must be added to the proof of (6) in [11], as it has been kindly communicated to the authors by P. Bénilan. See also the proof of (6) in [14]). With respect to the other free boundary $\zeta_1(t)$, its behaviour is given in the following theorem.

THEOREM 4. *There exists* $K > 0$ *and* $C > 0$ *such that*

$$Ct-K < \zeta_1(t) < +\infty \quad \text{for every } t > 0. \tag{7}$$

Moreover $\zeta_1(t)\uparrow+\infty$ *when* $t \to \infty$.

The second assertion of the above theorem is consequence of the principle of conservation of mass. Estimate (7) is obtained by comparing the solution u with a supersolution $\overline{u}$ of the form

$$\overline{u}(x,t) = \begin{cases} [f(x-Ct+K)]^{1/(m-\lambda)} & \text{if } x \geq Ct-K \\ 0 & \text{if } x < Ct-K \end{cases}$$

for a suitably chosen function f [11].

REMARK 4. Many of the above results can be established for some more general equations like (5), other initial-boundary value problems or in higher space dimension (see, for instance, [16], [12] and [9]).

4. An application to certain conservation laws equations

As a byproduct of Theorem 3, we can study the domain of dependence of the conservation law equation

$$u_t + f(u)_x = 0 \quad \text{in} \quad Q \tag{8}$$

when f is a continuous but not locally Lipschitz continuous function with $f(0) = 0$.

As is well-known ([20],[19],[7] and [2]), if f is locally Lipschitz continuous and $u_0 \in L^1(R) \cap L^\infty(R)$, then

$$\text{supp } u_0 = [a_1,a_2] \text{ implies supp } u(t,.) \subset [a_1-Kt,\ a_2+Kt],$$

where

$$K = \text{Sup } \{|f'(r)|, -\|u_0\|_\infty \leq r \leq \|u_0\|_\infty\}.$$

The local Lipschitz continuity assumed for f may be replaced by an integral condition near the origin. In particular, in [13] it is shown that supp u(t,.) is a compact set of $(-\infty,\infty)$ for every fixed $t \geq 0$, provided that f satisfies the condition

$$\int_0 \frac{ds}{|f^{-1}(s)|} < +\infty.$$

Nevertheless, both assumptions fail when f near the origin behaves like the function $f(s) = s^\lambda$, with $0 < \lambda < 1$. In fact, an explicit example due to Kruskov and Hildebrant shows that in that case the support of u(t,.) may be unbounded for every $t > 0$. The following result shows that this property is peculiar to a class of nonlinear functions f.

<u>THEOREM 5.</u> *Let* $u_0 \in L^1(R) \cap L^\infty(R)$, $u_0 \geq 0$, *with* supp $u_0 = [a_1,a_2]$. *Let* f *be a continuous function such that*

$$\int_0 \frac{ds}{|f(s)|} < +\infty. \tag{9}$$

Then if u *is the unique (entropy) solution of* (CP), *there exists a function* $\zeta_1(t)$ *such that* $u(t,x) > 0$ *for* $x > \zeta_1(t)$.

The idea of the proof consists in applying Theorem 3 - or, more precisely, a generalization of it - to the solutions of the equation

$$u_t - \varepsilon u_{xx} + f(u)_x = 0, \quad \epsilon > 0. \tag{10}$$

Indeed, as noted in Remark 4, the conclusion of Theorem 3 is true in a more general context, which includes, in particular, the case of equation (10) under the assumption (9) on the convection term [12]. Finally, by well-known results (see, e.g., [4]), lim $u_\varepsilon = u$ when $\varepsilon \downarrow 0$, and the conclusion is obtained by means of some uniform estimates on $\zeta_{1,\varepsilon}(t)$ [4].

<u>REMARK 5.</u> A different proof of Theorem 5 for the special case of $f(s) = |s|^{\lambda-1}s$, with $0 < \lambda < 1$, may be obtained via the estimate $u_t \geq -u/t$ proven in [8]. On the other hand, we point out that a result similar to Theorem 5 is also available for suitable N-dimensional conservation laws equations.

REFERENCES

1. Aronson, D.G.: Regularity properties of flows through porous media, SIAM J. Appl. Math. <u>17</u> (1969), 461-467.
2. Bénilan, P.: Equations d'evolution dans un espace de Banach quelconque et applications, Thèse, Orsay (1972).
3. Bénilan, P.: Evolution Equations and Accretive Operators, Lecture Notes, University of Kentucky (1981).
4. Bénilan, P. and H. Touré: Sur L'equation générale $u_t = \varphi(u)_{xx} - \Psi(u)_x + v$, C.R. Acad. Sci. Paris <u>299</u> (1984), 919-922.
5. Bertsch, M. and D. Hilhorst: A density dependent diffusion equation in population dynamics: stabilization to equilibrium, SIAM J. Math. Anal. (to appear).
6. Bertsch, M., R. Kersner and L.A. Peletier: Positivity versus localization in degenerate diffusion problems, Nonlinear Anal. TMA <u>9</u> (1985), 987-1008.
7. Crandall, M.G.: The semigroup approach to first order quasilinear equations in several space variables, Israel J. Math. <u>93</u> (1971), 265-298.
8. Crandall, M.G. and M. Pierre: Regularizing effects for $u_t + A\varphi(u) = 0$ in L^1, J. Funct. Anal. <u>45</u> (1982), 194-212.
9. Diaz, J.I.: Nonlinear Partial Differential Equations and Free Boundaries II: Parabolic and Hyperbolic Equations (Pitman, to appear).
10. Diaz, J.I. and R. Kersner: On a nonlinear degenerate parabolic equation in infiltration or evaporation through a porous medium (MRC Technical Summary Report # 2502, Univ. of Wisconsin-Madison, 1981).
11. Diaz, J.I. and R. Kersner: Non existence d'une des frontières libres dans une équation dégénerée on théorie de la filtration, C.R. Acad. Sci. Paris <u>296</u> (1983), 505-508.
12. Diaz, J.I. and R. Kersner: On the unboundedness of the domain of dependence on the initial data for some conservation laws equations (to appear).
13. Diaz, J.I. and L. Veron: Existence Theory and Qualitative Properties of the Solution of Some First Order Quasilinear Variational Inequalities, Indiana Univ. Math. <u>32</u> (1983), 319-361.

14. Francis, C.: On the porous medium equation with lower order singular nonlinear terms (to appear).

15. Gilding, B.H.: Properties of solutions of an equation in the theory of filtration, Arch. Rational Mech. Anal. 65 (1977), 203-225.

16. Gilding, B.H.: A nonlinear degenerate parabolic equation, Ann. Scuola Norm. Sup. Pisa 4 (1977), 393-432.

17. Gilding, B.H. and L.A. Peletier: The Cauchy problem for an equation in the theory of infiltration, Arch. Rational Mech. Anal. 61 (1976), 127-140.

18. Kalashnikov, A.S.: On the character of the propagation of perturbations in processes described by quasilinear degenerate parabolic equations, Proceedings of the seminar dedicated to I.G. Petrovskogo pp. 135-144 (1975).

19. Kruskov, S.N.: First order quasilinear equations in several independent variables, Math. USSR-Sb. 10 (1970), 217-243.

20. Lax, P.D.: Hyperbolic systems of conservations laws and the mathematical theory of shock waves. CBMS Regional Conference Series in Applied Mathematics 11 (SIAM, 1973).

21. Oleinik,O.A., A.S. Kalashnikov and C. Yui-Lin: The Cauchy problem and boundary value problems for equations of the type of nonstationary filtration, Izv. Akad. Nauk SSSR Sci. Mat. 22 (1958), 667-704.

22. Wu, Dequan: Uniqueness of the weak solution of quasilinear degenerate parabolic equations, Acta Math. Sinica 25 (1982), 61-75.

23. Wu, Zhuoqun and Junning Zhao: The first boundary value problem for quasilinear degenerate parabolic equations of second order in several space variables, Chinese Ann. Math. 4B (1983), 57-76.

24. Wolanski, N.I.: Flow through a Porous Column, Math. Anal. Appl. 109 (1985), 140-159.

J.I. Diaz
Departamento de Ecuaciones Funcionales
Facultad de Matemáticas
Universidad Complutense
28040 Madrid
España

R. Kersner
Magyar Tudományos Akadémia
Számitástechnikai és Automatizálási
Kutató Intézete
P.O. Box 63
H-1502 Budapest
Hungary

A FASANO

A free boundary problem in combustion

1. Introduction

Most of the material presented here (namely the existence, uniqueness and continuous dependence theorems for Problem (P) below) is a condensed form of a joint paper with J.R. Cannon and J.C. Cavendish [1]. A section on special solutions is added. We consider an idealized model for the combustion of a half-space full of solid fuel by a half-space full of gaseous oxidizer.

Neglecting heat conduction in the solid, evaporation of the solid and compressibility and convection in the gas, we are led to the following problem.

<u>PROBLEM (P).</u> For any given $T > 0$, find a curve $x = s(t)$ and two functions $u(x,t)$, $v(x,t)$ such that

(i) $s \in C^1(0,t] \cap C[0,t]$,

(ii) u, v are continuous and bounded in $-\infty < x \leq s(t)$, $0 \leq t \leq T$,

u_x, v_x are continuous in $-\infty < x \leq s(t)$, $0 < t \leq T$,

u_t, v_t, u_{xx}, v_{xx} are continuous in $-\infty < x < s(t)$, $0 < t \leq T$,

(iii) the following system is satisfied:

$$u_t = \alpha u_{xx}, \quad v_t = \beta v_{xx} \text{ in } -\infty < x < s(t),\ 0 < t \leq T, \tag{1.1}$$

$$s(0) = 0, \tag{1.2}$$

$$u(x,0) = \varphi(x), \quad v(x,0) = \psi(x), \ -\infty < x < 0, \tag{1.3}$$

$$\alpha u_x(s(t),t) = -(\gamma + u(s(t),t))\,\dot{s}(t), \quad 0 < t \leq T, \tag{1.4}$$

$$\beta v_x(s(t),t) = -(-\mu + v(s(t),t))\,\dot{s}(t), \ 0 < t \leq T, \tag{1.5}$$

$$\dot{s}(t) = \nu f(u(s(t),t)) \exp\{-\delta/(v(s(t),t) + v_o)\}, \quad 0 < t \leq T, \tag{1.6}$$

where:

(a) $\alpha, \beta, \gamma, \delta, \mu, \nu, v_0$ are positive given constants,

(b) φ, ψ are continuous on $-\infty < x \leq 0$, and

$$0 < \varphi(x) \leq \varphi^*, \qquad 0 < \psi_* \leq \psi(x) \leq \psi^*, \tag{1.7}$$

(c) f is continuously differentiable on $-\infty < u < +\infty$ and

$$f'(u) > 0 \quad \text{for} \quad u \neq 0, \tag{1.8}$$

$$f(0) = 0 . \tag{1.9}$$

In the scheme above all variables are nondimensional; u is related to the oxidizer concentration, v to temperature. Equations (1.4), (1.5) say that some of the oxidizer is consumed and some heat is released at the free boundary, while equation (1.6) describes the reaction rate.

See [1] for further references.

2. A priori estimates

Using the strong maximum principle, the so-called boundary point principle (or Hopf's lemma) together with the boundary conditions (1.4), (1.5), (1.6), and taking into account (a) - (c) it can be seen first that

$$0 < u(x,t) \leq \varphi^*, \quad \text{for} \quad -\infty < x \leq s(t), \quad 0 \leq t \leq T, \tag{2.1}$$

and then that

$$v_* = \min(\mu, \psi_*) \leq v(x,t) \leq \max(\mu, \psi^*). \tag{2.2}$$

As an obvious consequence

$$0 < \dot{s}(t) \leq \nu f(\varphi^*). \tag{2.3}$$

3. Continuous dependence upon the data

<u>THEOREM 3.1.</u> *If* (φ_i, ψ_i), $i = 1,2$ *are initial data, both satisfying assumption* (b), *and if* (u^i, v^i, s_i) *denote the corresponding solutions of* Problem (P), *then there exist two positive constants* C_1, C_2, *depending on the constants appearing in* (a) - (b), *on* sup f(u), sup f'(u) *for* $u \in (0, \varphi^*)$, *and on* T, *such that*

$$\text{(i)} \quad |s_1(t) - s_2(t)| + |\dot{s}_1(t) - \dot{s}_2(t)| \leq C_1\{\|\varphi_1 - \varphi_2\| + \|\psi_1 - \psi_2\|\}$$

($\|\cdot\|$ *being the* L^∞ norm); *moreover,*

(ii) $|u^1(x,t) - u^2(x,t)| + |v^1(x,t) - v^2(x,t)| \leq C_2\{\|\varphi_1 - \varphi_2\| + \|\psi_1 - \psi_2\|\}$

at any point where both u^1, v^1 *and* u^2, v^2 *are defined.*

Proof. The proof is rather long and will be summarized as follows.

Introducing the transformation

$$\xi = x - s(t),$$
$$U(\xi,t) = u(\xi+s(t),t),\ V(\xi,t) = v(\xi+s(t),t),$$

(1.1) - (1.6) become

$$U_t = \alpha U_{\xi\xi} + \dot{s}(t)U_\xi,\quad V_t = \beta V_{\xi\xi} + \dot{s}(t)V_\xi,\ -\infty < \xi < 0,\ 0 < t \leq T, \tag{3.1}$$

$$s(0) = 0, \tag{3.2}$$

$$U(\xi,0) = \varphi(\xi),\ V(\xi,0) = \psi(\xi),\ -\infty < \xi < 0, \tag{3.3}$$

$$\alpha U_\xi(0,t) = -(\gamma+U(0,t))\,\dot{s}(t),\quad 0 < t \leq T, \tag{3.4}$$

$$\beta V_\xi(0,t) = -(-\mu+V(0,t))\,\dot{s}(t),\ 0 < t \leq T, \tag{3.5}$$

$$\dot{s}(t) = \nu f(U(0,t))\exp\{-\delta/(V(0,t) + v_0)\},\quad 0 < t \leq T. \tag{3.6}$$

The analysis of the continuous dependence of s, U, V on the data φ, ψ is based on the classical representation formulas:

$$U(\xi,t) = \int_{-\infty}^{0} N(\xi,\eta,\alpha t)\varphi(\eta)d\eta \tag{3.7}$$

$$- 2\int_0^t K(\xi,\alpha(t-\tau))(\gamma+U(0,\tau))\dot{s}(\tau)\,d\tau$$

$$+ \alpha\int_0^t\int_{-\infty}^{0} N(\xi,\eta,\alpha(t-\tau))\,\dot{s}(\tau)\,U_\xi(\eta,\tau)\,d\eta\,d\eta\ ,$$

$$U_\xi(\xi,t) = \int_{-\infty}^{0} N_\xi(\xi,\eta,\alpha t)\,\varphi(\eta)d\eta \tag{3.8}$$

$$- 2\int_{-\infty}^{0} K_\xi(\xi,\alpha(t-\tau))(\gamma+U(0,\tau))\,\dot{s}(\tau)\,d\tau$$

$$+ \alpha\int_0^t \int_{-\infty}^0 N_\xi(\xi,\eta,\alpha(t-\tau))\ \dot{s}(\tau)U_\xi(\eta,\tau)d\eta\ d\eta$$

and the corresponding ones for V, where

$$K(x,t) = (4\pi t)^{-1/2}\ \exp\left\{-\frac{x^2}{4t}\right\}, \quad t > 0$$

and

$$N(\xi,\eta,t) = K(\xi-\eta,t) + K(\xi+\eta,t), \quad t > 0.$$

For any pair of solutions (s_i, U^i, V^i), $i = 1,2$ from (3.6) we derive the inequality

$$|\dot{s}_1(t) - \dot{s}_2(t)| \leq C\{\|U^1 - U^2\|_t + \|V^1 - V^2\|_t\}, \tag{3.9}$$

where $\|F\|_t = \sup\limits_{\substack{-\infty<\xi<0\\0<\tau<t}} |F(\xi,\tau)|$ and C denotes any constant depending on the same quantities as C_1, C_2.

Next, (3.8) and the estimates obtained in Section 2 give

$$\|U_\xi^i(\cdot,t)\| \leq C(1+t^{-1/2}),\ t > 0,\quad i = 1,2, \tag{3.10}$$

which can in turn be used along with a Gronwall type lemma to get

$$\|(U_\xi^1 - U_\xi^2)(\cdot,t)\| \leq C(1+t^{-1/2})\ \|\varphi_1 - \varphi_2\| + C\ \{\|U^1 - U^2\|_t + \|V^1 - V^2\|_t\}. \tag{3.11}$$

Clearly, similar inequalities hold replacing U by V and φ by ψ.

Using all the estimates above in (3.7) we obtain an integral inequality to which a generalized Gronwall lemma can be applied, implying

$$\|U^1 - U^2\|_t + \|V^1 - V^2\|_t \leq C\{\|\varphi_1 - \varphi_2 + \|\psi_1 - \psi_2\|\}. \tag{3.12}$$

Going back to the original variables and recalling (3.9), we obtain the estimates stated in the theorem.

4. Existence

<u>THEOREM 4.1.</u> *Under assumptions* (a) - (c) *there exists a solution to* Problem (P).

<u>Proof.</u> Let us make use of a retarded argument technique. A family of approximations $(s^\varepsilon, u^\varepsilon, v^\varepsilon)$, $\varepsilon > 0$, is obtained replacing u in (1.6) by

$$\overline{u}^\varepsilon(t) = \begin{cases} \varphi(0), & 0 \le t \le \varepsilon \\ u^\varepsilon(s^\varepsilon(t-\varepsilon), t-\varepsilon), & \varepsilon \le t \le T, \end{cases}$$

and v by $\overline{v}^\varepsilon(t)$ similarly defined.

Thus the triple $(s^\varepsilon, u^\varepsilon, v^\varepsilon)$ can be determined recursively, using such modified form of (1.6) to define s^ε at each time step. It is easily seen that all estimates found in Section 2 apply also to $(s^\varepsilon, u^\varepsilon, v^\varepsilon)$.

Suitable compactness results are obtained by proving the equicontinuity of $U^\varepsilon(0,t)$, $V^\varepsilon(0,t)$. This is done relating $U^\varepsilon(0,t_2) - U^\varepsilon(0,t_1)$ to the difference $U^\varepsilon(-d,t_2) - U^\varepsilon(-d,t_1)$ for some $d > 0$ and then using (3.10) and interior Schauder estimates with $t_2 \ge t_1 \ge \rho > 0$ (ρ being arbitrarily chosen). We get

$$|U^\varepsilon(0,t_2) - U^\varepsilon(0,t_1)| \le C^{(\rho)}(d + (t_2 - t_1)d^{-2}), \tag{4.1}$$

where $C^{(\rho)}$ depends on ρ. Taking $d = (t_2 - t_1)^{1/3}$ yields

$$|U^\varepsilon(0,t_2) - U^\varepsilon(0,t_1)| \le C^{(\rho)} (t_2 - t_1)^{1/3}. \tag{4.2}$$

The analysis of $V^\varepsilon(0,t)$ is similar.

Now we can select subsequences of $(s^\varepsilon, U^\varepsilon, V^\varepsilon)$ such that s^ε, $\dot{s}^\varepsilon$, $U^\varepsilon(0,t)$, $V^\varepsilon(0,t)$ converge uniformly to s, $\dot{s}$, U(0,t), V(0,t) on $\rho \le t \le T$ and U^ε, U^ε_ξ, U^ε_t; V^ε, V^ε_ξ, V^ε_t converge uniformly to U, U_ξ, U_t, V, V_ξ, V_t for $-\infty < \xi \le -\rho$, $\rho \le t \le T$. Moreover $s^{(\varepsilon)}$ converges uniformly in [0,T].

Passing to the limit in (3.7), written for U^ε and in the corresponding representation for V^ε, one sees that (U,V) satisfies (3.1) - (3.5) with s(t) representing the limit of $s^\varepsilon(t)$.

Finally, passing to the limit in

$$\dot{s}^{\varepsilon}(t) = \nu f(\overline{u^{\varepsilon}}(t)) \exp \{-\delta/(\overline{v^{\varepsilon}}(t) + v_o)\}$$

the correct free boundary condition is obtained for the triple (s,u,v).

5. Special solutions

An equivalent formulation of Problem (P) is obtained by means of the transformation

$$Z(x,t) = -\int_x^{s(t)} u(\xi,t)d\xi + \gamma(x - s(t)), \tag{5.1}$$

$$W(x,t) = -\int_x^{s(t)} u(\xi,t)d\xi - \mu(x - s(t)). \tag{5.2}$$

In particular, the free boundary conditons (1.4) - (1.6) become

$$Z(s(t),t) = W(s(t),t) = 0, \qquad 0 \leq t \leq T, \tag{5.3}$$

$$\dot{s}(t) = \nu f(Z_x - \gamma) \exp \{-\delta/(W_x + \mu + v_o)\} . \tag{5.4}$$

In this formulation many explicit solutions can be calculated, particularly in the class of processes exhibiting constant temperature on the combustion surface.

Taking e.g. $f(u) = f_o u^p$, f_o, $p > 0$, we can look for the solution corresponding to some given constant temperature on the free boundary and to the following penetration law for the combustion front

$$s(t) = s_o (1 - e^{-\omega t})^{1+Np}, \tag{5.5}$$

with s_o, ω positive constants and $N \in \{0,1,2,...\}$.

The solution can be obtained in the form of absolutely and uniformly convergent series. It can be proved that

$$Z(x,t) - \gamma(x-s(t)) = -\int_x^{s(t)} u(\xi,t)d\xi$$

tends to zero as $t \to +\infty$, uniformly with respect to x, i.e. the corresponding combustion process leaves no residual mass of the oxidizer. This result is not obvious in general for the problem in the half-space.

REFERENCES

1. Cannon, J.R., J.C. Cavendish and A. Fasano: A free boundary-value problem related to the combustion of a solid, SIAM J. Appl. Math. 45 (1985), 798-809.

A. Fasano
Istituto Matematico "U. Dini"
Università di Firenze
Viale Morgagni 67A
50134 Firenze
Italia

A FRIEDMAN

Optimal control for variational inequalities

Optimal control problems for variational inequalities are studied in [1], [5], [6] and in some of the references given there, and necessary conditions are obtained. In this paper* we wish to report on recent work ([2],[3]) whereby bang-bang principles are established for the optimal control; see also [4],[5].

Consider the variational inequality

$$\begin{aligned} -\Delta u &\geq -(f + k) \\ u &\geq 0 \qquad\qquad \text{a.e. in } \Omega \\ (-\Delta u + f+k)u &= 0 \\ u &= U^o \quad \text{on} \quad \partial\Omega \end{aligned} \tag{1}$$

where Ω is a bounded domain in R^N with smooth boundary, U^o is smooth and positive, f is a given function in $L^p(\Omega)$, $p \geq 2$, $p > N/2$, and k belongs to a set A:

$$A \text{ is a bounded, closed convex set in } L^p(\Omega). \tag{2}$$

We associate with k and the solution u of (1) a functional

$$J(k) = \int_\Omega F(x,u(x))dx \tag{3}$$

where F and F_u are continuous. One can easily show that there exists a $k_o \in A$ such that

$$J(k_o) = \max_{k\in A} J(k). \tag{4}$$

Our interest is to find the structure of any maximizer k_o. Since the functional J(k) is not differentiable, we shall approximate problem (4) by

*This work is partially supported by National Science Foundation Grant MCS-8300293.

smooth ε-problems as follows (cf. [1]):

Let u_ε be the solution of

$$-\Delta u + \beta_\varepsilon(u) = f + k \text{ in } \Omega, \quad u = U^0 \text{ on } \partial\Omega \tag{5}$$

where β_ε is the usual penalty function, and set

$$J_\varepsilon(k) = \int_\Omega F(x,u_\varepsilon) - \frac{1}{2}\int_\Omega (k - k_0)^2.$$

If $(k_\varepsilon, u_\varepsilon)$ is a maximizer of J_ε in A, then (by [1]) $k_\varepsilon \to k_0$ in $L^p(\Omega)$.

LEMMA 1. *Denote by Q_ε the solution of*

$$-\Delta Q_\varepsilon = F_u(x,u_\varepsilon) \quad \text{in } \Omega, \quad Q_\varepsilon = 0 \quad \text{on } \partial\Omega .$$

If $k_\varepsilon + \delta\ell \in A$ for any $0 < \delta < 1$, then

$$\int Q_\varepsilon \ell \geq 0.$$

From this lemma one can infer properties of k_ε. Suppose for instance that

$$A = \{0 \leq k \leq M, \int_\Omega k = H\}. \tag{6}$$

LEMMA 2. *If A is given by (6) then there exists a λ_ε such that*

$$k_\varepsilon = \begin{cases} 0 & \text{if} \quad Q_\varepsilon > \lambda_\varepsilon \\ M & \text{if} \quad Q_\varepsilon < \lambda_\varepsilon. \end{cases}$$

From Lemmas 1 and 2 one can deduce:

THEOREM 1. *If (k_0,u_0) is a solution of problem (4) then there exists a λ such that*

$$k_o = \begin{cases} 0 & \text{if } Q > \lambda \\ M & \text{if } Q < \lambda . \end{cases}$$

Here Q *is a solution of*

$$-\Delta Q = F_u(x,u_o(x)) \quad \text{in} \quad \Omega_o = \{u_o > 0\}$$

and $Q = 0$ *on* $\partial\Omega$.

THEOREM 2. *If* $F_u > 0$ *then* $\lambda \geq 0$, *and* $\{Q = \lambda\}$ *has measure zero; further, if* $\lambda > 0$ *then* $k_o = 0$ *in* $\{u_o = 0\}$. *Finally, if* $f \leq 0$ *then* $\lambda > 0$.

Theorems 1 and 2 exhibit the bang-bang nature of the optimal control k_o.

The above results can be extended to parabolic variational inequalities, to other control sets A and functionals J, and to problems in which the control occurs on the boundary. In particular, an application to the one-phase Stefan problem can be given in case the control function k is the flux: $\partial\vartheta/\partial\nu = k$ where ϑ is the temperature. The problem is to choose k

$(0 \leq k \leq M, \int_0^T\int_S k = M)$ on the fixed boundary S of the water in such a way so as to melt as much ice as possible. Extending Theorems 1 and 2 one can show that

$$k_o(x,t) = \begin{cases} M & \text{if } \quad 0 < t < \varphi(x) \\ 0 & \text{if } \quad \varphi(x) < t < T \end{cases}$$

where $\varphi(x)$ is some continuous function.

REFERENCES

1. Barbu, V.: Optimal Control of Variational Inequalities (Pitman, 1984).

2. Friedman, A.: Optimal control for variational inequalities, SIAM J. Control Optim. (to appear).

3. Friedman, A.: Optimal control for parabolic variational inequalities (to appear).

4. Friedman, A. and L. Jiang: Nonlinear optimal control problems in heat conduction, SIAM J. Control Optim. 21 (1983), 940-952.

5. Friedman, A. and D. Yaniro: Optimal control for the dam problem, Appl. Math. Optim. (to appear).

6. Mignot, F. and J.P. Puel: Optimal control in some variational inequalities, SIAM J. Control Optim. 22 (1984), 466-476.

7. Mignot, F. and J.P. Puel: Contrôl optimal d'un système gouverné par une inéquation variationelle parabolique, C.R. Acad. Sci. Paris 298 (1984), 277-280.

A. Friedman
Department of Mathematics
Northwestern University
Evanston, IL 60201
USA

M G GARRONI

Green's function and asymptotic behaviour of the solution of some oblique derivative problems not in divergence form

Introduction

The purpose of this paper is to present some new results about the asymptotic behaviour of the solutions of some elliptic and parabolic oblique derivative problems for an operator A not in divergence form and for a boundary operator B with Hölder continuous coefficients.

We consider the following parabolic problem

$$\begin{cases} \frac{\partial z}{\partial t} + Az = 0 & \text{on } \Omega \times]0,T[\\ Bz = 0 & \text{in } \partial\Omega \times [0,T] \\ z(\cdot,0) = \Phi & \text{on } \Omega. \end{cases} \tag{1}$$

We are interested in studying the asymptotic behaviour of the solution $z(x,t)$ when t goes to infinity and in showing how a "good" asymptotic behaviour of z allows to solve some non coercive stationary problems.

In particular, this is used to describe the limiting behaviour of the solutions v_λ of the unilateral problem:

$$\begin{cases} \max[v_\lambda \,;\, Av_\lambda + \lambda v_\lambda - f] = 0 & \text{in } \Omega \\ Bv_\lambda = 0 & \text{on } \partial\Omega \end{cases} \tag{2}$$

as the positive parameter λ tends to zero. This is motivated by the control theory of stochastic processes. The above boundary value problem actually encompasses the Bellman conditions for the optimal stopping of a diffusion process in Ω reflected at the boundary (see [2]). The case $\lambda > 0$ corresponds to discounted cost functionals while $\lambda = 0$ is related to long run average costs (see [9]).

This problem has been studied by A. Bensoussan and J.L. Lions (see [3]) in the case of regular coefficients. Their main result is that the asymptotic behaviour of u_λ, as λ tends to zero, depends on the sign of the mean value of f with respect to the measure m dx, m being the solution of the adjoint homogeneous problem

$$\begin{cases} A^* m = 0 \quad \text{in} \quad \Omega \\ B^* m = 0 \quad \text{on} \quad \partial\Omega,\ m > 0,\ \int_\Omega m\,dx = 1,\ m \in C^2(\overline{\Omega}). \end{cases}$$

In the general framework considered in this paper the main difficulties arise from the non-divergence structure of A. Actually, the fairly low regularity of the coefficients does not allow a Fredholm alternative approach using integration by parts. The key to solve these problems is the Green's function $G(x,y,t,\tau)$ for the initial boundary value problem associated with A and B, as constructed in ([7]).

1. Preliminary results

Let Ω be a bounded open subset of $\mathbb{R}^N$, $N \geq 2$, with boundary $\partial\Omega$ of class C^2. We shall denote by Q_T the cylinder $\Omega \times]0,t[$, $0 < \underline{T} < +\infty$ and by $\Sigma_T = \partial\Omega \times]0,T[$ its lateral boundary. We set

$$A = -\sum_{i,j=1}^{N} a_{ij}(x) \frac{\partial^2}{\partial x_i \partial x_j} + \sum_{i=1}^{N} a_i(x) \frac{\partial}{\partial x_i},\quad B = \sum_{i=1}^{N} b_i(x) \frac{\partial}{\partial x_i}, \tag{3}$$

whose coefficients are assumed to satisfy

$$\left| \begin{array}{ll} \text{(i)} & a_{ij}, a_i, b_i \in C^{0,\alpha}(\Omega), \text{ for some } 0 < \alpha < 1 \\ \text{(ii)} & \sum_{i,j=1}^{N} a_{ij}(x)\xi_i\xi_j \geq \mu|\xi|^2,\ \mu > 0,\ \forall \xi \in \mathbb{R}^N,\ \forall x \in \Omega \\ \text{(iii)} & \left|\sum_{i=1}^{N} b_i(x)\nu_i(x)\right| \geq \beta > 0,\ \forall x \in \partial\Omega . \end{array} \right. \tag{4}$$

Here $\nu = (\nu_1,\nu_2,\ldots,\nu_N)$ is the outward normal vector to $\partial\Omega$.

<u>THEOREM 1.</u> *Under the above assumptions there exists the Green's function* $G \equiv G(x,y,t,\tau)$ *of the initial boundary value problem:*

$$\begin{cases} \frac{\partial G}{\partial t} + A_x G = \partial(x-y)\partial(t-\tau), & \text{in } Q_T \\ G(x,y,t,\tau)|_{t\leq\tau} = 0 \\ B_x G = 0, \text{ on } \Sigma_T ; \end{cases} \tag{5}$$

G *satisfies the estimates:*

$$\begin{cases} |G(x,y,t,\tau)| \leq c_1(t-\tau)^{-\frac{N}{2}} e^{-c_2 \frac{|x-y|^2}{t-\tau}} \\ |\frac{\partial}{\partial x_i} G(x,y,t,\tau)| \leq c_1(t-\tau)^{-\frac{N+1}{2}} e^{-c_2 \frac{|x-y|^2}{t-\tau}} \\ |\frac{\partial}{\partial t^s} \frac{\partial}{\partial x^r} G(x,y,t,\tau)| \leq c_1(t-\tau)^{-\frac{N+1+\alpha}{2}} \rho^{\alpha-1}(x) V(t-\tau)^{\frac{\alpha-1}{2}} e^{-c_2 \frac{|x-y|^2}{t-\tau}}, \end{cases} \tag{6}$$

where $2s+r = 2$; $s,r \in \mathbb{N}$ *and* $\rho(x) = dist(x,\partial\Omega)$; *moreover we have:*

$$\begin{cases} \text{(i)} \quad G(x,y,t,\tau) \geq 0, \\ \text{(ii)} \quad \text{for every ball } B \subset \Omega \text{ and } t > \tau, \text{ there exists a constant } \delta \\ \qquad \text{such that } G(x,y,t,\tau) \geq \delta > 0 \quad \forall x \in \overline{\Omega}, \quad \forall y \in B, \\ \text{(iii)} \quad \int_\Omega G(x,y,t,\tau)dy = 1. \end{cases} \tag{7}$$

Results (5) and (6) are contained in a more general setting in [7]; result (7) is contained in [6].

Set $\varphi \in W_p^{2-2/p}(\Omega)$, $1 < p < \frac{1}{1-\alpha}$, satisfying convenient compatibility conditions (see [9] or [7]); from Theorem 1 it follows that the solution of problem 1 is given by

$$z(x,t) = \int_\Omega G(x,y,t,0)\Phi(y)dy. \tag{8}$$

THEOREM 2. *For every* $\lambda > 0$, *and* $f \in L^p(\Omega)$, $(1 < p < \frac{1}{1-\alpha})$, $\alpha < \frac{N-1}{N}$, *the oblique derivative problems:*

$$\begin{cases} u_\lambda \in W_p^2(\Omega) \\ Au_\lambda + \lambda u_\lambda = f \text{ in } \Omega, \; Bu_\lambda = 0 \text{ on } \partial\Omega; \end{cases} \tag{9}$$

and

$$\begin{cases} v_\lambda \in W^2_p(\Omega),\ \text{Max}[v_\lambda, Av_\lambda + \lambda v_\lambda - f] = 0 \text{ in } \Omega \\ Bv_\lambda = 0 \text{ on } \partial\Omega \end{cases} \tag{10}$$

have unique solution u_λ *and* v_λ *respectively; we have also the representation*

$$u_\lambda(x) \equiv (\Gamma_\lambda f)(x) = \int_0^{+\infty} e^{-\lambda t}[\int_\Omega G(s,y,t,0)f(y)dy]dt \tag{11}$$

(*see* [6]).

Theorem 2 extends to Hölder continuous boundary coefficients and exponents $1 < p < 2$ the results of [1], [3], [2] and [11].

2. Fredholm alternative, invariant measure and asymptotic behaviour.

We consider the following problems:

$$u \in L^p(\Omega),\ (I-\lambda\Gamma_\lambda)u = \Gamma_\lambda f \tag{12}$$

$$m \in L^{p'}(\Omega),\ (I-\lambda\Gamma^*_\lambda)m = 0 \quad \frac{1}{p} + \frac{1}{p'} = 1, \tag{13}$$

where Γ_λ is defined by (11) and Γ^*_λ is its adjoint.

<u>THEOREM 3.</u> *For every* $\lambda > 0$ *the equation* (13) *has a unique solution* m *such that*

$$m > 0,\ \frac{1}{|\Omega|}\int m dx = 1. \tag{14}$$

Moreover,

$$\int_\Omega [\int_\Omega G(x,y,t,0)f(y)dy]m(x)dx = \int_\Omega f(x)m(x)dx,\ \forall t > 0,\ \forall f \in L_p(\Omega), \tag{15}$$

$$\|\int_\Omega G(x,y,t,0)f(y)dy - \frac{1}{|\Omega|}\int fmdy\|_{L^p} \le k\|f\|_{L^p} e^{-\nu t},\ \forall t > 0,\ \forall f \in L^p(\Omega) \tag{16}$$

with k,ν *constants independent of* f *(see* [6]*).*

Since when f belongs to $W_p^{2-2/p}(\Omega)$, the integral $\int_\Omega G(x,y,t,0)f(y)dy$ satisfies problem 1(see (8)). Then property (15) says that the average of the solution $z(x,t)$ is constant for every positive t, and property (16) gives the asymptotic behaviour of the solution $z(x,t)$ when t goes to infinity.

Theorem 3 has a probabilistic interpretation. Actually, we can define probability transition functions P by setting

$$P(x,t,E) = \int_E G(x,y,t,0)dy,$$

for any Borel subset E of $\overline{\Omega}$.

These are related to a diffusion process reflected at the boundary $\partial\Omega$ according to the vector field B, and $d\mu = \frac{1}{|\Omega|}\, m dx$ is the invariant measure of this process (see [6] for details).

The behaviour of the solution u_λ and v_λ and the form of the limiting problems depend on $\int_\Omega fmdx$, where m is given by Theorem 3. Conditions (15) and (16) are essential to prove the following theorem (see [6]):

<u>THEOREM 4.</u> *Let* $f \in L^p$ $(1 < p < \frac{1}{1-\alpha})$, $\alpha > \frac{N-1}{N}$; *then* $\|u_\lambda\|_{W^2_p}$ *is bounded by a constant independent of* λ *if and only if* $\int_\Omega fmdx = 0$. *In this case*

$\lim_{\lambda\to 0} u_\lambda = u$ *in* W^2_p *weakly and*

$$u(x) = \int_0^\infty \left(\int_\Omega G(x,y,t,0)f(y)dy\right)dt \qquad (17)$$

is the unique solution such that $\int_\Omega umdx = 0$.

(i) *If* $\int_\Omega fmdx > 0$, v_λ *tends weakly in* $W^2_p(\Omega)$ *to the unique solution* v *of the problem*

$$v \in W^2_p(\Omega),\ \mathrm{Max}[v;Av-f] = 0 \text{ in } \Omega,\ Bv = 0 \text{ on } \partial\Omega. \qquad (18)$$

(ii) *If* $\int_\Omega fmdx < 0$, *then* $w_\lambda = v_\lambda - \frac{1}{|\Omega|}\int_\Omega v_\lambda mdx$ *tends weakly in* $W^2_p(\Omega)$ *to the unique solution* w *of the problem*

$$\begin{cases} w \in W_p^2(\Omega), \ Aw = f - \frac{1}{|\Omega|}\int_\Omega f m dx \ \textit{in} \ \Omega, \ Bw = 0 \ \textit{on} \ \partial\Omega, \\ \int_\Omega w m dx = 0. \end{cases} \tag{19}$$

(iii) *If* $\int fmdx = 0$, $p > \frac{N}{2}$, *then* v_λ *tends weakly in* $W_p^2(\Omega)$ *to the maximum solution* v *of the problem*

$$v \in W_p^2(\Omega), \ v \leq 0, \ Av = f \ \textit{in} \ \Omega, \ Bv = 0 \ \textit{on} \ \partial\Omega. \tag{20}$$

These results can be extended to a certain integro-differential operator A+J (A as in (3)), where J was introduced in [2]), corresponding to a reflected diffusion process with interior jumps. We can also add a non linear operator H depending on the gradient, more precisely an Hamiltonian operator.

This has been done in a joint work with J.L. Menaldi [8].

REFERENCES

1. Agmon, S., A. Douglis and L. Nirenberg: Estimates near the boundary for solutions of elliptic P.D.E.'s Comm. Pure Appl. Math. 12 (1959), 623-727.
2. Bensoussan, A. and J.L. Lions: Contrôle impulsionnel et inequations quasi-variationnelles. (Dunod, 1982).
3. Bensoussan, A. and J.L. Lions: On the asymptotic behaviour of the solution of variational inequalities. In "Summer School on the Theory of Non-linear Operators", pp. 25-40, Akademie Verlag, 1978).
4. Chicco, M.: Third boundary value problem in $H^{2,p}(\Omega)$ for a class of linear second order elliptic partial differential equations, Rend. Istit. Mat. Univ. Trieste 4 (1972), 80-94.
5. Capuzzo-Dolcetta, I. and M.G. Garroni: Comportement asymptotique de la solution de problèmes non sous forme divergence avec condition de derivée oblique sur le bord, C.R. Acad. Sci. Paris 299 (1984), 843-846.
6. Capuzzo-Dolcetta, I. and M.G. Garroni: Oblique derivative problems and invariant measures, Ann. Scuola Norm. Sup. Pisa (to appear).
7. Garroni, M.G. and V.A. Solonnikov: On parabolic oblique derivative problems with Hölder continuous coefficients. Comm. Partial Differential Equations 9 (1984), 1323-1372.
8. Garroni, M.G. and J.L. Menaldi: Green's function and invariant measures for integro-differential problems (in preparation).
9. Robin, M.: On some impulse control problems with long run average cost, SIAM J. Control Optim. 19 (1981), 333-358.

10. Solonnikov, V.A.: On boundary value problems for linear general parabolic systems of differential equations, Trudy Mat. Inst. Steklov 83 (1965).

11. Troianiello, G.M.: Book to appear.

Further references on the subject can be found in the papers [6] and [7].

M.G. Garroni
Dipartimento di Matematica,
Università di Roma "LA SAPIENZA"
Piazza A. Moro
00185 Roma
Italia

M A HERRERO

Temperature fronts in one dimensional media

1. Description of the problem; solutions with interfaces

In this note we survey some recent results concerning the Cauchy problem:

$$u_t - (u^m)_{xx} + \lambda u^n = 0 \qquad (-\infty < x < +\infty, \quad t > 0) \tag{1.1}$$

$$u(x,0) = u_0(x) \qquad (-\infty < x < +\infty). \tag{1.2}$$

The following assumptions will be retained henceforth:

u_0 is a continuous nonnegative function, which vanishes outside a finite interval $I = (a,b)$ $(-\infty < a < b < +\infty)$; (H1)

$m > 1$, $\lambda > 0$, $n > 0$. (H2)

Equation (1.1) is usually referred to as the one-dimensional nonlinear heat equation with absorption. In this model u stands for the temperature, $(u^m)_{xx}$ is the diffusion term and (λu^n) represents the volumetric absorption of heat by the medium. It is apparent that both the thermal conductivity (mu^{m-1}) and the absorption coefficient (λu^{n-1}) are temperature-dependent: these facts have a deep influence on the behaviour of solutions.

When $\lambda = 0$ equation (1.1) goes into the nonlinear heat conduction or porous media equation:

$$u_t = (u^m)_{xx} \qquad (-\infty < x < +\infty, \quad t > 0, \ m > 1), \tag{1.3}$$

which has been extensively dealt with in recent years (see for instance [22] for a survey up to 1979). The most striking property of (1.3) is the onset of temperature waves or solutions with interfaces. Namely, under assumption (H1) the Cauchy problem has a unique solution (to be defined in a generalized way), which is positive within an expanding set bounded by two interfaces (free boundaries or fronts) $x = \zeta_i(t)$ $(i = 1,2)$, which are defined for $t \geq 0$

as follows:

$$\zeta_1(t) = \inf\{x \in \mathbb{R} : u(x,t) > 0\}, \ \zeta_2(t) = \sup\{x \in \mathbb{R} : u(x,t) > 0\} \tag{1.4}$$

(see [21]). Physically, fronts originate since the conductivity (mu^{m-1}) vanishes when $u = 0$, which happens at the boundary of the support of u_o at $t = 0$. This certainly does not occur in the constant-conductivity case $m = 1$, where perturbations propagate with infinite speed, nor for (1.3) when u_o is positive everywhere, in which case solutions remain positive (see [3]).

The porous media application of (1.3) motivates the following widespread terminology (see [1]). The unknown $u = u(x,t)$ stands for the density of the gas and

$$v = \frac{m}{m-1} u^{m-1} \quad \text{is called the \underline{pressure};} \tag{1.5}$$

$$V = -v_x = -mu^{m-2}\, u_x \text{ is the \underline{local velocity of the gas}} \tag{1.6}$$

(defined in the set where $u \not\equiv 0$).

We shall find it convenient to keep to this notation in the absorption case $\lambda > 0$.

An example of solutions of (1.3) which exhibit interfaces are the Barenblatt functions (see [4]). Using the pressure defined in (1.5) they read as follows:

$$\overline{v}(x,t) \equiv \overline{v}(x,t;M) = (2(m+1)t)^{-1}[r^2(t)-x^2]_+ , \tag{1.7}$$

where $[s]_+ = \max\{s,0\}$ and:

$$\begin{cases} r(t) = c_m\, M^{\frac{m-1}{m+1}} \cdot t^{\frac{1}{m+1}} , \\ c_m = \left\{\dfrac{2m(m+1)}{m-1}\; B\left(\dfrac{m}{m-1}, \dfrac{1}{2}\right)^{1-m}\right\}^{\frac{1}{m+1}} , \end{cases} \tag{1.8}$$

$B(\cdot,\cdot)$ being the Euler beta function. It is easy to see that $\overline{u} = \left(\frac{m-1}{m}\,\overline{v}\right)^{1/(m-1)}$ satisfies (1.3) in $D'(\mathbb{R} \times (0,\infty))$, but not in the classical sense (it fails to do so precisely at the fronts $x = \zeta_i(t) = \pm r(t)$). On the other hand, as $t \downarrow 0$ one has:

$$\bar{u}(x,t) \to M\delta(x), \tag{1.9}$$

where $\delta(x)$ is the Dirac delta centered at the origin. Solutions of PDE's like (1.1) or (1.3) satisfying (1.9) are often called <u>source-type solutions.</u> They play a key role in many problems as models describing the large-time behaviour of solutions corresponding to arbitrary initial data (see for instance [5]). For a description of the asymptotics of (1.3) see [12], [24]).

When $\lambda > 0$, one can obtain source-type solutions of (1.1) in some cases by modifying Barenblatt functions in a suitable way. For instance, if <u>$m+n = 2$</u>, the quantity:

$$\tilde{v}(x,t) \equiv \tilde{v}(x,t;M) = (2(m+1)t)^{-1}[r^2(t) - x^2 - \lambda(m+1)^2t^2]_+, \tag{1.10}$$

where $M > 0$ and $r(t)$ is given in (1.8), is such that $\tilde{u} = \left(\frac{m-1}{m}\tilde{v}\right)^{\frac{1}{m-1}}$ is a source-type solution of (1.1) (see [19], [15], where a time-shifted version of $\tilde{v}$, $\tilde{v}_\tau(x,t) = \tilde{v}(x,t+\tau)$, $\tau > 0$ was first derived). It is easy to check that $\tilde{u}$ reduces to $\bar{u}$ above if $\lambda = 0$.

As for (1.10), the interfaces are now given by

$$(\tilde{\zeta}_i(t))^2 = c_m^2\, M^{\frac{2(m-1)}{m+1}} \cdot t^{\frac{2}{m+1}} - \lambda(n+1)^2t^2 \quad (i = 1,2), \tag{1.11}$$

c_m being defined in (1.8). Thus, for instance, the right-hand interface moves forward (<u>heating wave</u>) if $t < T_R$, where

$$T_R = \left[\frac{c_m^2}{\lambda(m+1)^3}\right]^{\frac{m+1}{2m}} \cdot M^{\frac{m-1}{m}} .$$

It then recedes (<u>cooling wave</u>) for $t > T_R$ and collapses to the origin at $t = T_E = (m+1)^{(m+1)/2m} \cdot T_R$; in such case the whole solution disappears. T_E is usually called the <u>extinction time</u> of $\tilde{u}$.

When <u>$n = 1$</u> it is known [20] that the change of variables:

$$\begin{cases} \tau = \dfrac{1}{\lambda(m-1)} (1 - \exp(-\lambda(m-1)t)) \\ u(x,t) = \hat{u}(x,\tau)e^{-\lambda t} \end{cases} \tag{1.12}$$

maps (1.1) into (1.3) in the sense that, if $u(x,t)$ solves (1.1) with $n = 1$, $\hat{u}(x,\tau)$ satisfies

$$\hat{u}_\tau = (\hat{u}^m)_{xx}.$$

We then obtain from (1.7), (1.12) a further source-type solution, which in the pressure variable reads:

$$v(x,t) = e^{-\lambda(m-1)t} \left(\frac{2(m+1)}{\lambda(m-1)} (1 - e^{-\lambda(m-1)t})\right) \times \tag{1.13}$$

$$\times \left[c_m^2 M^{\frac{2(m-1)}{m+1}} \left[\frac{1 - e^{-\lambda(m-1)t}}{\lambda(m-1)}\right]^{\frac{2}{m+1}} - x^2\right]_+ .$$

Note that now the interfaces are uniformly bounded for any $t > 0$, and therefore heat propagates only into a finite region of space.

Source-type solutions are known to appear when $n < m + 2$, but they cannot exist if $n > m + 2$([13]). No other explicit example is available by now. However, another class of solutions has been recently obtained in [6] for the case $n = m$. They are given by:

$$y(x,t) = \left[\frac{1}{f(t)} \left[\rho - \frac{ch(\alpha x) - 1}{g(t)}\right]_+\right]^{\frac{1}{m-1}} . \tag{1.14}$$

Here $\rho > 0$ is arbitrary, $\alpha = \sqrt{\lambda}\left(\frac{m-1}{m}\right)$ and f, g are positive increasing functions that satisfy:

$$\int_A^{g(t)} [s(s + \frac{2}{\rho})]^{\frac{m-1}{2}} \, ds = ct,$$

$$f(t) = \lambda(m-1)\, \rho t + \lambda \frac{(m-1)^2}{m} \int_0^t \frac{ds}{g(s)} + t_o ,$$

where A, c, t_o are nonnegative constants. Note that the interfaces satisfy now:

$$|\zeta_i(t)| \to \infty \quad \text{as} \quad t \to \infty \quad (i = 1,2).$$

To end this section we remark that the previously given solutions do not satisfy (1.1) in the classical sense at the fronts. As in the case $\lambda = 0$, this shows that solutions of (1.1), (1.2) have to be considered in a generalized way. The corresponding theory has been fairly developed by now and can be found for instance in [11], [16], [9].

2. The behaviour of the fronts

Going back to equation (1.3), one can show by physical considerations [2] that when fronts occur, they have to move with the local velocity of the gas near them. This leads to the equation:

$$\zeta_i'(t) = V(\zeta_i(t),t) \equiv \lim_{\substack{x\to\zeta_i(t)\\ |x|<|\zeta_i(t)|}} (-v_x(x,t)) \quad (i = 1,2) \tag{2.1}$$

(see [2], [17], [7]), which is certainly satisfied by the Barenblatt functions (see (1.7), (1.8)).

When $\lambda > 0$ the natural question arises of determining the influence of the absorption on the evolution of the fronts (which always exist under our hypotheses; see [11]). This influence is likely to be all the more important when $n < 1$, in which case the absorption coefficient (λu^{n-1}) goes to infinity as u approaches the fronts. Actually, in this case solutions with interfaces are known to exist even in the constant-conductivity case:

$$u_t - u_{xx} + \lambda u^n = 0 \qquad (-\infty < x < +\infty, \quad t > 0). \tag{2.2}$$

Moreover, fronts can appear in (2.2) for $t > 0$, even if the initial data u_o is positive everywhere (see for instance [8]).

To better understand the effect of the absorption, we now borrow from [23] to formally expand the solution in a power series near a moving interface. More precisely, let (x_o,t_o) be a point on, say, the left-hand interface (i.e., $x_o = \zeta_1(t_o)$); let us try in (1.1):

$$u(x,t) = k(x-x_o + A(t-t_o))^{\ell} + k_1(x-x_o + A(t-t_o)^{\ell_1} + \dots , \tag{2.3}$$

where $x - x_o > 0$, $t - t_o > 0$ and $A = -\zeta_i'(t_o)$ is the velocity of the front. Retaining only the lowest relevant powers, we obtain:

$$Ak\ell(x-x_o+ A(t-t_o)^{\ell-1}+\dots = k^m\, m\ell(m\ell-1)(x-x_o+ A(t-t_o))^{m\ell-2} - \tag{2.4}$$
$$- k^n(x-x_o+ A(t-t_o)^{n\ell} + \dots .$$

Then, as detailed in [23], the following possibilities appear:

(i) $\ell-1 = mn-2$, $n\ell > \ell-1$. In this case $\ell = \frac{1}{m-1}$ and $m+n > 2$. From (2.4) it follows that, if A is to be positive,

$$\zeta_i'(t_o) = V(\zeta_1(t_o),t_o) \tag{2.5}$$

which is the same equation as for the case $\lambda = 0$.

(ii) $\ell-1 = n\ell < m\ell-2$. Then $\ell = \frac{1}{1-n}$ for $n < 1$ and $m+n > 2$. We then formally deduce from (2.4) that for negative A:

$$\zeta_i'(t_o) = W(\zeta_1(t_o),t_o), \tag{2.6}$$

where:

$$W(\zeta_1(t_o),t_o) = \lim_{\substack{x\to\zeta_1(t_o)\\ x>\zeta_1(t_o)}} \left(-(1-n)\left(\frac{\partial}{\partial x}\left(u^{1-n}\right)\right)^{-1}\right).$$

These facts strongly suggest that, if $m+n > 2$ and the front expands, its velocity should be given by (2.5), whereas if it shrinks (which never happens for (1.3): see [3]; or for (1.1) with $n \geq 1$: see [14], [18] the corresponding expression should be that in (2.6).

(iii) $m\ell-2 = n\ell = n-1$. Then $m+n = 2$, and

$$\zeta_i'(t_o) = V(\zeta_1(t_o),t_o) - \lambda m(V(\zeta_1(t_o),t_o))^{-1} . \tag{2.7}$$

The analysis of the case m+n < 2 is more involved and will be omitted here.

It is now easy to check that (2.5) is actually satisfied by the explicit solutions (1.13), (1.14), while (2.7) holds for the source-type solution corresponding to m+n = 2 (see (1.10)). Therefore the question of the rigorous justification of (2.5) - (2.7) arises. We merely sketch here how to do it in the case $n \geq m$ [9]. To begin with, it is known that whenever $m+n \geq 2$ the function

$$v_x = \frac{m}{m-1} \frac{\partial}{\partial x}(u^{m-1})$$

is bounded in any strip $S_\tau = \mathbb{R} \times (\tau,\infty)$ with a bound depending on τ and $\sup_{x\in\mathbb{R}} v(x,\tau)$ ([10], [9]; (this is no longer true if m+n < 2; see [10]). This means that under the above assumptions the quantity defined in (1.6) is bounded; however, in order to get, say, (2.5) we have yet to show that the limit in the right-hand side exists. On the other hand, when $n \geq m$ one has

$$v_{xx} \geq -\frac{k}{t} \text{ in } D'$$

for $t > 0$ and some $k = k(m,n) > 0$, so that for fixed positive time the quantity

$$v_x(x,t) + \frac{kx}{t}$$

has lateral limits at any point $x \in \mathbb{R}$. Then the rest of the proof is obtained by means of comparison with suitable sub- and supersolution.

Finally, justification of (2.6), (2.7) is currently in progress.

REFERENCES

1. Aronson, D.G.: SIAM J. Appl. Math. 17 (1969), 461-467.
2. Aronson, D.G.: Arch. Rational Mech. Anal. 37 (1970), 1-10.
3. Aronson, D.G. and P. Bénilan: C.R. Acad. Sci. Paris 288 (1979), 103-105.
4. Barenblatt, G.I.: Prikl. Mat. Meh. 16 (1952), 67-68.
5. Barenblatt, G.I.: Similarity, self-similarity and intermediate asymptotics (Plenum Press, 1981).
6. Bertsch, M., R. Kersner and L.A. Peletier: Nonlinear Anal. TMA 9 (1985), 987-1008.

7. Caffarelli, L.A. and A. Friedman: Amer. J. Math. 101 (1979), 1193-1218.
8. Evans, L.C. and B.F. Knerr: Illinois J. Math. 23 (1979), 153-166.
9. Herrero, M.A. and J.L. Vazquez: SIAM J. Math. Anal. (to appear).
10. Kalashnikov, A.S.: Moscow Univ. Math. Bull. 33 (1978), 44-51.
11. Kalashnikov, A.S.: USSR Computational Math. and Math. Phys. 16 (1976), 689-696.
12. Kamin, S.: Israel J. Math. 14 (1973), 76-78.
13. Kamin, S. and L.A. Peletier: (to appear).
14. Kersner, R.: Acta Math. Acad. Sci. Hungar. 34 (1979), 157-163.
15. Kersner, R.: Moscow Univ. Math. Bull. 33 (1978), 44-51.
16. Kersner, R.: Nonlinear Anal. TMA 4 (1980), 1043-1062.
17. Knerr, B.F.: Trans. Amer. Math. Soc. 234 (1977), 381-415.
18. Knerr, B.F.: Trans. Amer. Math. Soc. 249 (1979), 409-424.
19. Martinson, L.K.: J. Appl. Mech. Tech. Phys. 21 (1980), 419-421.
20. Martinson, L.K. and K.B. Pavlov: USSR Computational Math. and Math. Phys. 12 (1979), 261-268.
21. Oleinik, O.A., A.S. Kalashnikov and C.Y. Lin: Izv. Akad. Nauk SSSR Ser. Mat. 22 (1958), 667-704.
22. Peletier, L.A.: The porous media equation. In "Applications of Nonlinear Analysis in the Physical Sciences", H. Amann, N. Bazley and K. Kirchgässner eds., pp. 229-241 (Pitman, 1981).
23. Rosenau, P. and S. Kamin: Phys. D8 (1983), 273-283.
24. Vazquez, J.L.: Trans. Amer. Math. Soc. 277 (1983), 507-527.

M.A. Herrero
Departamento de Ecuaciones Funcionales
Facultad de Matemáticas
Universidad Complutense
28040 Madrid
España

P HESS

Positive solutions of periodic-parabolic problems and stability

1. In this note we study the question of existence of positive periodic solutions (of given period $T > 0$) of semilinear parabolic problems of the form

$$\begin{cases} Lu = g(x,t,u) & \text{on} \quad \Omega \times \mathbb{R} \\ Bu = 0 & \text{on} \quad \partial\Omega \times \mathbb{R} \\ u(\cdot,t) = u(\cdot,t+T) & \text{on} \quad \overline{\Omega}, \ \forall t \in \mathbb{R}, \end{cases} \tag{1}$$

where $L := \frac{\partial}{\partial t} + A(x,t,\frac{\partial}{\partial x})$ is a uniformly parabolic linear differential expression of second order having T-periodic coefficient functions, $g : \overline{\Omega} \times \mathbb{R} \times \mathbb{R} \to \mathbb{R}$ is T-periodic in t, and Ω is a bounded domain in $\mathbb{R}^N$ $(N \geq 1)$. Further $B = B(x,\frac{\partial}{\partial x})$ is a linear boundary operator implying Dirichlet, Neumann or regular oblique derivative bondary conditions. Problems of type (1) arise naturally e.g. in population dynamics, if one looks at the population density in a non-homogeneous medium and assumes that both diffusion and growth rate are subject to seasonal variations.

It is well-known that periodic-parabolic problems behave in many respects like elliptic boundary value problems. Searching for <u>positive</u> solutions, as in the elliptic theory the notions of <u>principal eigenvalue</u> and <u>eigenfunction</u> are of particular usefulness, for the construction of positive sub- and supersolutions, for the application of bifurcation arguments, for stability considerations. The linear eigenvalue problem

$$\begin{cases} Lu = \lambda m(x,t)u & \text{in} \quad \Omega \times \mathbb{R} \\ Bu = 0 & \text{on} \quad \partial\Omega \times \mathbb{R} \\ u(\cdot,t) = u(\cdot,t+T) & \text{on} \quad \overline{\Omega}, \ \forall t \in \mathbb{R}, \end{cases} \tag{2}$$

has first been investigated for $m = \mathbb{1}$ and $Bu = u$ by Lazer [9] and Castro-Lazer [3]; for a general T-periodic weight function m Beltramo-Hess [2] and Hess [6] have given a both necessary and sufficient condition for the existence of a positive principal eigenvalue, assuming that (L,B) satisfies the maximum principle. Here we extend these results to more general operators $\mathcal{B}$ and study in particular also the resonance case where 0 is a principal eigenvalue, searching for a second, nontrivial, principal eigenvalue (results by Beltramo [1], extending the corresponding elliptic results by Hess and Senn [14,7]). These results have obvious immediate applications to bifurcation theory: one gets a both necessary and sufficient condition for the bifurcation of a continuum of positive periodic solutions from a line of trivial solutions.

As an example we study the following model equation in population genetics, a spatially inhomogeneous periodic version of Fisher's equation

$$\begin{cases} \frac{\partial u}{\partial t} - a(x,t)\Delta u = s(x,t)h(u) & \text{in } \Omega \times \mathbb{R} \\ \frac{\partial u}{\partial n} = 0 & \text{on } \partial\Omega \times \mathbb{R} \\ u(\cdot,t) = u(\cdot,t+T) & \text{on } \overline{\Omega},\ \forall t \in \mathbb{R}\,. \end{cases} \tag{3}$$

Here a and s are smooth and T-periodic in t, a is positive on $\overline{\Omega} \times \mathbb{R}$, and n denotes the outer normal to $\partial\Omega$. It is assumed that $h \in C^1(\mathbb{R})$ satisfies $h(0) = h(1) = 0$, $h(u) > 0$ for $0 < u < 1$, and $h'(0) > 0$, $h'(1) < 0$, and that s changes sign in $\overline{\Omega} \times [0,T]$. (3) admits the trivial solutions $u = 0$ and $u = \mathbb{1}$; only u with $0 \leq u \leq \mathbb{1}$ are of interest. We give sufficient conditions for the existence of nontrivial periodic solutions u: $0 < u < \mathbb{1}$ and, applying a new result of Dancer-Hess [4], show that (3) always admist at least one stable periodic solution. (For a discussion of the existence of nontrivial equilibrium solutions if $a = \mathbb{1}$ and s is independent of t, see [5, 12, 13].)

2. In the bounded domain $\Omega \subset \mathbb{R}^N$ with boundary $\partial\Omega$ of class $C^{2+\mu}$ $(0 < \mu < 1)$, let $L := \frac{\partial}{\partial t} + A(x,t,\frac{\partial}{\partial x})$ be a uniformly parabolic linear differential expression with

$$A(x,t,\frac{\partial}{\partial x})u = -\sum_{j,k=1}^{N} a_{jk}(x,t)\frac{\partial^2 u}{\partial x_j \partial x_k} + \sum_{j=1}^{N} a_j(x,t)\frac{\partial u}{\partial x_j} + a_o(x,t)u.$$

We assume that, for fixed $T > 0$, the coefficient functions $a_{jk} = a_{kj}$, a_j, a_o belong to the real Banach space $E := \{w \in C^{\mu,\mu/2}(\overline{\Omega} \times \mathbb{R}) : w \text{ is T-periodic in } t\}$. Further let $\beta \in C^{1+\mu}(\partial\Omega, \mathbb{R}^N)$ be an outward pointing, nowhere tangent vector field on $\partial\Omega$ and $b \in C^{1+\mu}(\partial\Omega)$, $b \geq 0$. Define the boundary operator $\mathcal{B}$

either by $\mathcal{B}u := u$ (implying Dirichlet boundary conditions) or by $\mathcal{B}u := \frac{\partial u}{\partial \beta} + bu$ (implying Neumann or regular oblique derivative boundary conditions). Let $L : E \supset D(L) \to E$ be the operator in E induced by $\mathcal{L}$, $\mathcal{B}$ and the periodicity condition. As in [2,6] one proves that $D(L) := F := \{w \in C^{2+\mu, 1+\mu/2}(\overline{\Omega} \times \mathbb{R}) : \mathcal{B}w = 0$ on $\partial\Omega \times \mathbb{R}$ and w is T-periodic in t$\}$, and that L is a closed operator with compact resolvent. Moreover, since for sufficiently large $c > 0$ $(\mathcal{L}+c, \mathcal{B})$ satisfies the maximum principle, $(L+c)^{-1} : E \to F$ is strongly positive. By the Krein-Rutman theorem $(L+c)^{-1}$ hence has a unique (simple) positive eigenvalue having a positive eigenfunction u. Thus there is a unique principal eigenvalue $\alpha_1 \in \mathbb{R}$ of

$$Lu = \alpha_1 u \quad (u > 0).$$

Let $U_c(t,s)$ denote the fundamental solution associated with L+c (more precisely: with the parabolic equation

$$\frac{du}{dt}(t) + (A(t) + c)u(t) = f(t),$$

where $A(t) = \mathcal{A}(\cdot, t, \frac{\partial}{\partial x}))$; $K_c := U_c(T,0)$ is then a compact positive operator in $C := C(\overline{\Omega})$ having spectral radius $0 < \gamma_c := spr(K_c) < 1$, and $\alpha_1 = -\frac{1}{T} \log \gamma_c - c$.

$(\mathcal{L}, \mathcal{B})$ satisfies the maximum principle if and only if $\alpha_1 > 0$.

Let $m \in E$ be a given weight function, and let $M \in \mathcal{L}(E)$ denote the associated multiplication operator. In order to treat the general eigenvalue problem (2) it is important to study the one-parameter family of eigenvalue problems

$$(L-\lambda M)u = \alpha u \qquad (\lambda \in \mathbb{R}).$$

By the above, to each $\lambda \in \mathbb{R}$ there is a unique eigenvalue $\alpha = \alpha_1(\lambda) \in \mathbb{R}$ having a positive eigenfunction $u_1 = u_1(\lambda) \in F$. As a consequence of the implicit function theorem, $\alpha_1(\lambda)$ is an analytic function of λ, and also the mapping $\lambda \to u_1(\lambda) \in F$ can be chosen to be analytic. It follows either from Kato [8], or from a direct argument as in [2], that $\alpha_1(\lambda)$ is a concave function of λ.

We introduce the quantities

$$P(m) := \int_0^T (\max_{x\in\overline{\Omega}} m(x,t))dt, \quad N(m) := \int_0^T (\min_{x\in\overline{\Omega}} m(x,t))dt.$$

If $m \in E$ is independent of $x \in \Omega$, i.e. $N(m) = P(m)$, $\alpha_1(\lambda)$ can be calculated directly as

$$\alpha_1(\lambda) = \alpha_1(0) - \frac{\lambda}{T} P(m).$$

For general m we have (cf. [2,6])

$$P(m) > 0 \Rightarrow \lim_{\lambda\to+\infty} \alpha_1(\lambda) = -\infty,$$

$$N(m) < 0 \Rightarrow \lim_{\lambda\to-\infty} \alpha_1(\lambda) = -\infty.$$

On the other hand, estimates from below can be obtained by a simple comparison argument:

$$\alpha_1(\lambda) \geq \alpha_1(0) - \frac{\lambda}{T} P(m) \qquad \text{for } \lambda > 0,$$

$$\alpha_1(\lambda) \geq \alpha_1(0) - \frac{\lambda}{T} N(m) \qquad \text{for } \lambda < 0,$$

with strict inequalities if m depends nontrivially on $x \in \Omega$.

From these auxiliary results it follows that if $(L,\mathcal{B})$ satisfies the maximum principle, a necessary and sufficient condition for the existence of a positive principal eigenvalue of (2) is that $P(m) > 0$ (note that $\alpha_1(0) > 0$).

3. We now focus attention to the <u>resonance</u> <u>case</u> where the coefficient function of the 0th order term of A, $a_o = 0$, and where $\mathcal{B}u := \frac{\partial u}{\partial \beta}$; here $\alpha_1(0) = 0$ and $u_1(0) = \underline{1}$. For the eigenvalue problem (2) we then have

<u>THEOREM 1.</u>

A) *If* $N(m) = P(m) = 0$, *then* $\sigma(L,M) = \mathbb{C}$, *and each* $\lambda \in \mathbb{R}$ *is eigenvalue of* (2) *with a positive eigenfunction. In all the other cases the spectrum* $\sigma(L,M)$ *is discrete.*

B) *If* $N(m) = P(m) \neq 0$, 0 *is the only eigenvalue of* (2) *with a positive eigenfunction.*

C) *If* $N(m) < P(m)$ *and either* $P(m) \leq 0$ *or* $N(m) \geq 0$, 0 *is the only eigenvalue of* (2) *having a positive eigenfunction.*

D) *Let* $N(m) < 0 < P(m)$. *There is a real-valued linear functional* I *acting on* m, *such that*

(i) *if* $I(m) \neq 0$, *then* (2) *has a unique nontrivial eigenvalue* $\lambda_1(m)$ *with positive eigenfunction, and* $\lambda_1(m) \gtrless 0$ *if* $I(m) \lessgtr 0$. *Further* 0 *and* $\lambda_1(m)$ *are* M-*simple eigenvalues of* L.

(ii) *if* $I(m) = 0$, *then* 0 *is the only eigenvalue of* (2) *with a positive eigenfunction.*

(We recall that $\lambda \in \mathbb{C}$ is M-simple eigenvalue of L if $\dim N(L-\lambda M) = \operatorname{codim} R(L-\lambda M) = 1$ and $Mu \notin R(L-\lambda M)$, where $N(L-\lambda M) = \text{span}[u]$.)

The functional I is given by

$$I(m) = \left\langle \emptyset^*, \int_0^T U_1(t,\tau)e^{-\tau}m(\cdot,\tau)d\tau \right\rangle ,$$

where $\emptyset^* \in C^*$ is the unique positive eigenfunction to K_1^* (to the eigenvalue $\gamma_1 = e^{-T}$), normalized by $\|\emptyset^*\|_{C^*} = 1$.

Theorem 1 follows from the above auxiliary results if one notes that in cases A) and B) $\alpha_1(\lambda)$ is a linear function, and that

$$\frac{d\alpha_1}{d\lambda}(0) = -(Te^T \langle \emptyset^*, 1 \rangle)^{-1} \; I(m).$$

4. Let the function $g : \overline{\Omega} \times \mathbb{R} \times \mathbb{R} \times \mathbb{R}$ be such that $g(x,t,u)$ is T-periodic in t and of class $C^{\mu,\mu/2}(\overline{\Omega} \times \mathbb{R})$ in (x,t) uniformly for u in bounded intervals, and that $\frac{\partial g}{\partial u}$ is continuous on $\overline{\Omega} \times \mathbb{R} \times \mathbb{R}$.

Let $u \in F$ be a solution of (1). Lazer [9] has shown that the principle of linearized stability holds: setting $m(x,t) := \frac{\partial g}{\partial u}(x,t,u(x,t))$, u is locally asymptotically exponentially stable if $\alpha_1 > 0$ for the eigenvalue problem $(L-M)w = \alpha_1 w \; (w > 0)$, and unstable if $\alpha_1 < 0$.

The following assertion is a particular case of a result of Dancer-Hess [4] and generalizes work by Matano [10] on equilibrium solutions of autonomous problems.

PROPOSITION. *Let* $u_1 < u_2$ *be two order-related solutions of* (1). *If we restrict the concept of stability to the order-interval* $[u_1,u_2]$ *(i.e. consider only initial conditions in* $[u_1,u_2]/_{t=0} \subset C^{2+\mu}(\overline{\Omega})$), *there exists at least one stable solution of* (1) *in* $[u_1,u_2]$.

5. We now turn to the model problem (3) of selection-migration in population genetics. As a consequence of the proposition we have

THEOREM 2. *Problem* (3) *admits at least one solution* $u : 0 \leq u \leq \mathbb{1}$ *which is stable with respect to the order-interval* $[0,\mathbb{1}]$.

We can be more precise: for the nonlinearity $g(x,t,u) = s(x,t)h(u)$ set

$$m_o(x,t) := \frac{\partial g}{\partial u}(x,t,0) = h'(0)s(x,t),$$

$$m_1(x,t) := \frac{\partial g}{\partial u}(x,t,1) = h'(1)s(x,t),$$

and recall that $h'(0) > 0$, $h'(1) < 0$. Parallel to the statement of Theorem 1 we distinguish between 4 cases:

A) If $N(s) = P(s) = 0$ we are in an exceptional situation: searching for spatially constant functions $u = w(t)\mathbb{1}$ of (3), we reduce the problem to the ODE

$$\frac{dw}{dt}(t) = s(t)h(w(t)) \text{ on } \mathbb{R}\text{, } w \text{ is T-periodic.} \tag{4}$$

It is easily seen that for each initial condition $w_o \in]0,\mathbb{1}[$ for $t = 0$, the solution w of (4) is T-periodic; all the corresponding solutions $u = w\mathbb{1}$ are stable solutions of (3) by the comparison theorem.

B) If $N(s) = P(s) \neq 0$, either the trivial solution 0 or $\mathbb{1}$ is stable.

C) The same is true if $N(s) < P(s)$ and either $P(s) \leq 0$ or $N(s) \geq 0$.

D) Let $N(s) < 0 < P(s)$. Then

(i) if $I(s) < 0$, the solution 0 is stable if $\lambda_1(m_o) > 1$ and unstable if $\lambda_1(m_o) < 1$; the solution $\mathbb{1}$ is always unstable;

(ii) if $I(s) > 0$, the same assertion holds for the solutions 0 and $\mathbb{1}$ interchanged;

(iii) if $I(s) = 0$, both trivial solutions are unstable.

If both $\lambda_1(m_o) < 1$ and $\lambda_1(m_1) < 1$, we can construct strict (periodic) sub- and supersolutions $\underline{v} < \overline{v}$ in $]0,\mathbb{1}[$ (cf. [11, Theorem 4.2]). The proposition then guarantees the existence of a stable periodic solution in

$]\underline{v},\overline{v}[$.

In the situation (i) of D, looking at the nonlinear eigenvalue problem $Lu = \gamma s\, h(u)$ $(\gamma \in \mathbb{R})$ associated with (3), it follows from Theorem 1, D that we have global bifurcation of an unbounded continuum of positive T-periodic solutions (γ,u), from the line $\mathbb{R}\times\{0\}$ of trivial solutions, at $(\lambda_1(m_0),0)$. The continuum lies entirely in the strip $]0,\infty[\times]0,\underline{1}\![$ of $\mathbb{R}\times F$. If $h \in C^3$ and $h''(0) < 0$, we have bifurcation to the right, and the positive solutions are stable in a neighbourhood of the bifurcation point. A similar result holds for bifurcation from the line $\mathbb{R}\times\{\underline{1}\!\}$. (See [13] for a proof in the stationary case.)

REFERENCES

1. Beltramo, A.: Ueber den Haupteigenwert von periodisch-parabolischen Differentialoperatoren, Ph.D. Thesis, University of Zurich (1984).
2. Beltramo, A. and P. Hess: On the principal eigenvalue of a periodic-parabolic operator. Comm. Partial Differential Equations 9 (1984), 919-941.
3. Castro, A. and A.C. Lazer: Results on periodic solutions of parabolic equations suggested by elliptic theory. Boll. Un. Mat. Ital. IB (1982), 1089-1104.
4. Dancer, N. and P. Hess: (to appear).
5. Fleming, W.H.: A selection-migration model in population genetics, J. Math. Biol. 2 (1975), 219-234.
6. Hess, P.: On positive solutions of semilinear periodic-parabolic problems, Lecture Notes in Mathematics 1076, pp. 101-114 (Springer,1984).
7. Hess, P. and S. Senn: Another approach to elliptic eigenvalue problems with respect to indefinite weight functions, lecture Notes in Mathematics 1107, pp. 106-114 (Springer, 1985).
8. Kato, T.: Superconvexity of the spectral radius, and convexity of the spectral bound and type. Math. Z. 180 (1982), 265-273.
9. Lazer, A.C.: Some remarks on periodic solutions of parabolic differential equations. In "Dynamical Systems II", Bednarek-Cesari Eds., pp. 227-246 (Academic Press, 1982).
10. Matano, H: Existence of nontrivial unstable sets for equilibriums of strongly order-preserving systems, J.Fac. Sci. Univ. Tokyo 30 (1984), 645-673.
11. Sattinger, D.H.: Monotone methods in nonlinear elliptic and parabolic boundary value problems, Indiana Univ. Math. J. 21 (1972), 979-1000.
12. Saut, J.C. and B. Scheurer: Remarks on a nonlinear equation arising in population genetics. Comm. Partial Differential Equations 3 (1978), 907-931.

13. Senn, S.: On a nonlinear elliptic eigenvalue problem with Neumann boundary conditions, with an application to population genetics. Comm. Partial Differential Equations 8 (1983), 1199-1228.

14. Senn, S. and P. Hess: On positive solutions of a linear elliptic eigenvalue problem with Neumann boundary conditions, Math. Ann. 258 (1982), 459-470.

P. Hess
Mathematisches Institut
Universität Zürich
Rämistrasse 74
CH-8001 Zürich
Switzerland

S KAMIN

Asymptotic properties of the solutions of porous medium equations with absorption

We consider the Cauchy problem

$$u_t = \Delta u^m - u^p \quad \text{in} \quad \mathbb{R}^n \times (0,\infty) \tag{1}$$

$$u(x,0) = \varphi(x) \quad \text{in} \quad \mathbb{R}^n \tag{2}$$

in which $m \geq 1$, $p > 1$ and φ is a given bounded nonnegative function. In recent papers ([3],[5],[6]) the behaviour as $t \to \infty$ of the solutions $u(x,t)$ of (1), (2) has been studied. It was shown that the large time behaviour of u depends on the behaviour of $\varphi(x)$ for large x. To specify the behaviour of φ we introduce the parameter $\alpha > 0$ through the hypothesis

$$\lim_{|x|\to\infty} |x|^{\alpha} \varphi(x) = A, \tag{3}$$

where A is a positive number. Such asymptotic behaviour of u depends also on the competition between the diffusion and absorption term in (1), i.e. on the parameters m and p. It was proved for different cases that for large t the solution $u(x,t)$ is close to some similarity solution. Here we are interested in the case $\alpha = n$ which happened to be a borderline case: for $\alpha < n$ and $\alpha > n$ the large time behaviour of u is completely different. We show that if $\alpha = n$ for large t the solution $u(x,t)$ is close to some corrected similarity solution. We explain it below.

Let us consider first the Cauchy problem for the porous medium equation

$$w_t = \Delta w^m \tag{4}$$

$$w(x,0) = \varphi(x). \tag{5}$$

Suppose

$$\varphi(x)\ |x|^n \to A \quad \text{as} \quad |x| \to \infty\ . \tag{6}$$

As in [6] we use the stretching transformation. Let

$$w_k(x,t) = f(k)w(kx,k^2f^{m-1}(k)t). \tag{7}$$

It is readily seen that for every $k > 0$ and every function $f(k)$, w_k is a solution of (4) with initial data

$$w_k(x,0) = \varphi_k(x) = f(k)\varphi(kx).$$

Now we put

$$f(k) = \frac{1}{\int_{B_1} \varphi(k\xi)d\xi} \tag{8}$$

where B_1 is the unit ball $B_1 = \{x, |x| < 1\}$. With this choice of $f(k)$ we obtain

$$\int_{B_1} \varphi_k(x)dx = 1 \tag{9}$$

and, by (5)

$$\varphi_k(x) = f(k)\varphi(kx) = \frac{k^n\varphi(kx)}{\int_{B_k} \varphi(z)dz} \to 0 \tag{10}$$

as $k \to \infty$ for every $x \neq 0$. The equations (9) and (10) mean that

$$\varphi_k(x) \to \delta(x) \quad \text{as} \quad k \to \infty \quad \text{in} \quad D'.$$

Therefore we expect that

$$w_k(x,t) \to E(x,t) \quad \text{as} \quad k \to \infty. \tag{11}$$

Here $E(x,t)$ is the fundamental solution of the heat equation if $m = 1$ and the source-type solution of (4) if $m > 1$ ([2]). Now, suppose one could prove the convergence in (11). Then we have

$$f(k)w(kx,k^2f^{m-1}t) \to E(x,t) = k^nE(kx,k^{2+n(m-1)}t).$$

If we now set $t = 1$, $kx = y$, $\tau = k^{2+n(m-1)}$, we get

$$\lim_{\tau\to\infty} f(\tau^{1/2+n(m-1)})w(y,\tau^{2/2+n(m-1)}f^{m-1}(\tau^{1/2+n(m-1)})) \tag{12}$$
$$= \lim_{\tau\to\infty} \tau^{n/2+n(m-1)}E(y,\tau).$$

The relation (12) may also be written as

$$w(x,t) \sim \frac{\tau^{n/2+n(m-1)}}{f(\tau^{1/2+n(m-1)})} E(x,\tau)$$

where $t = \tau^{2/2+n(m-1)}f^{m-1}(\tau^{1/2+n(m-1)})$.

Supposing (6) we obtain for large k

$$f(k) \sim \frac{k^n}{AS_n \ell nk}$$

where S_n is the surface area of the unit sphere in $\mathbb{R}^n$. Therefore

$$w(x,t) \sim \frac{AS_n \ell n\tau}{2+n(m-1)} \; E(x,\tau)$$

where

$$t = \left[\frac{2+n(m-1)}{AS_n}\right]^{m-1} \frac{\tau}{(\ell n\tau)^{m-1}} \; .$$

The last equality means that for large time the solution $w(x,t)$ is close to the somewhat corrected similarity solution. The word corrected means here that the similarity solution E is taken at a different moment of time and also multiplied by some function. This correction is what makes this bordline case different from the cases $\alpha < n$ and $\alpha > n$. In the latter the asymptotic behaviour of w is given by appropriate pure similarity solutions ([1], [4]).

Let us mention the particular case $m = 1$. Then $t = \tau$ and

$$w(x,t) \sim \tfrac{1}{2}AS_n \ell nt \; E(x,t).$$

Here the difference between the cases $\alpha < n$ and $\alpha = n$ shows up explicitly.

For $\alpha < n$ the solution $w(x,t)$ is close to $E(x,t)$ in the sense that

$$t^{n/2}|w(x,t)-E(x,t)| \to 0, \quad t \to \infty$$

with

$$\int E(x,t)dx = \int \varphi(x)dx.$$

Correspondingly for $\alpha = n$ we find

$$t^{n/2}\left|\frac{2}{AS_n}\frac{w(x,t)}{\ell nt} - E(x,t)\right| \to 0$$

and

$$\int E(x,t)dx = 1.$$

The precise formulations and the proofs of the results mentioned above may be found in [7]. We shall also present in [7] some results when $\varphi(x)$ satisfies slightly more general conditions then (3). For instance we can handle the case

$$\varphi(x) \sim \frac{\ell n|x|}{|x|^{\alpha}} \quad \text{as} \quad |x| \to \infty .$$

The same kind of considerations may be used to deal with equation (1) if $\alpha = n$ and $p > m + \frac{2}{n}$. Using the same stretching transformation

$$u_k(x,t) = f(k)u(kx,k^2f^{m-1}(k)t)$$

with $f(k)$ given by (8), we find that u_k is the solution of the equation

$$u_t = \Delta u^m - F(k)u^p,$$

where $F(k) = k^2[f(k)]^{m-p}$. In this case $F(k) \to 0$ as $k \to \infty$ and it can be proved that

$$u_k(x,t) \to E(x,t) \quad \text{as} \quad k \to \infty . \tag{13}$$

As before the asymptotic behaviour for u(x,t) follows from (13). More details may be found in [7].

REFERENCES

1. Alikakos, N.D. and R. Rostamian: On the uniformization of the solutions of the porous medium equation in $\mathbb{R}^n$, Israël J. Math., 47 (1984), 270-290.
2. Barenblatt, G.I.: On some unsteady motions in a liquid or a gas in a porous medium, Prikl. Mat. Meh. 16 (1952), 67-78.
3. Escobedo, M. and O. Kavian: Asymptotic behaviour of positive solutions of a nonlinear heat equation (to appear).
4. Friedman, A. and S. Kamin: The asymptotic behaviour of gas in an n-dimensional porous medium, Trans. Amer. Math. Soc. 262 (1980), 551-563.
5. Gmira, A. and L. Veron: Large time behaviour of solutions of a semilinear equation in $\mathbb{R}^n$, J. Differential Equations 53 (1984), 258-276.
6. Kamin, S. and L.A. Peletier: Large time behaviour of solutions of the porous media equation with absorption (to appear).
7. Kamin, S. and M. Ughi (to appear).

S. Kamin
School of Mathematical Sciences
Tel-Aviv University
Ramat-Aviv 69978
Tel-Aviv
Israël

R KERSNER

Extinction in finite time versus non-extinction in nonlinear diffusion problems

1. It is known that the solution of the Cauchy problem

$$u_t = u_{xx} \qquad (t > 0,\ x \in \mathbb{R}) \tag{1}$$

$$u(x,0) = u_o(x) \qquad (x \in \mathbb{R}) \tag{2}$$

with nonnegative continuous initial values such that supp $u_o \subset [-\ell,\ell]$, $(\ell > 0)$ behaves as $1/\sqrt{t}$ when t is large (notation: $u(x,t) \sim 1/\sqrt{t}$; we will not be interested in exact asymptotics).

2. The simple change $u = ve^{c_o t}$ shows that with the same $u_o(x)$ for the equation

$$u_t = u_{xx} - c_o u, \qquad c_o = \text{const} > 0 \tag{3}$$

we have

$$u(x,t) \sim \frac{e^{-c_o t}}{\sqrt{t}}$$

3. It follows from the maximum principle that ($u_o(x)$ being the same) the solution of the equation

$$u_t = u_{xx} - c_o u^n \tag{4}$$

is bounded below by the solution of (3) if $n > 1$.

4. If $0 < n < 1$ in (4), then the situation becomes more complicated. It is necessary to introduce the notion of generalized solution ("solution in the sense of distributions"), see e.g. [5]. To the class of such solutions already belong functions like

$$v(x,t) = v(t) = [T - C_o(1-n)t]_+^{\frac{1}{1-n}}, \tag{5}$$

where $[f]_+ = \max(f,0)$.

It follows from (5) that $u = 0$ if $t \geq T_o = \overset{*}{T}(C_o(1-n))^{-1}$. In such a situation - there exists a $T_o > 0$ such that $u(x,t) = 0$ for $t \geq T_o$ - we shall speak about extinction in finite time (or localization in time according to another terminology).

5. When we are concerned with the more general equation

$$u_t = (u^m)_{xx} - c_o u^n, \quad m > 1, \quad n > 0, \tag{6}$$

the investigation of the behaviour of the solutions becomes more complicated due to the appearance of a so-called finite speed of propagation of disturbances: if the support of $u_o(x)$ is bounded, then the support of $u(x,t)$ will be bounded in x for $t > 0$ too.

6. In order to understand the transition between the various equations mentioned above ("$C_o \searrow 0$", "$n \nearrow 1$", "$n \searrow 1$", "$m \searrow 1$" and so on), we have to consider the more general equation

$$u_t = (\varphi(u))_{xx} - c(u) \tag{7}$$

where $\varphi(0) = 0$, $\varphi'(0) \geq 0$; $\varphi, \varphi', \varphi'' > 0$ for $u > 0$,
$c(0) = 0$; $c, c' > 0$ for $u > 0$.

The fact that the problem (7), (2) is well-posed (existence, uniqueness, comparison theorems, regularity) is proved in [6].

Let

$$P = \int_{0^+} \frac{ds}{c(s)}$$

It is known [5] that if $P < \infty$ then, in (7), (2), total extinction occurs in finite time.

The proof of this follows from the fact that, in this case, the solution can be estimated from above by a generalized solution analogous to (5). The suitable v(t) can be defined as follows where $T = \int_0^{\max u_o} \frac{ds}{c(s)}$:

$$\int_0^{v(t)} \frac{ds}{c(s)} = T - t \text{ for } t < T,$$

$$v(t) = 0 \text{ for } t \geq T.$$

One can see from this proof that a similar phenomenon can appear for equations of a very different form. Acutally, we only used the fact that $\varphi(v)_{xx} = 0$, the only important property being that the corresponding equation satisfies a "maximum principle". For example, the equation

$$u_t = \sum_{i,j=1}^{N} \frac{\partial}{\partial x_i} (a_{ij}(t,x,u,\text{grad } u)) - c(u)$$

could be even "very" degenerated - a_{ij} could vanish on the whole interval - while at the same time extinction in finite time occurs. For the first-order equation

$$u_t + g(u)u_x + \lambda u^{\alpha} = 0, \quad \lambda > 0, \quad 0 < \alpha < 1$$

similar results were first obtained in [9].

We note that localization in time can exist not only because of strong absorption but also because of strong diffusion, see [1], [2], [3], [4], [11] and the references given there.

7. The following question comes out in a natural way: is the condition $P < \infty$ necessary too? For the Cauchy problem (7), (2) the following should be true, although it has not been proved:

$$\text{if } P = \infty \text{ and } u_o(x_o) > 0, \text{ then } u(x_o,t) > 0 \text{ for } t > 0.$$

There are some results in this direction: for $\varphi(u) = u$ in [10]; for the general case see Theorems 1,2.

From the formal point of view the proof of this conjecture would solve the question. But actually a complete answer should require further details. The question is that non-extinction can take place in various ways, depending on the kind of diffusion (i.e. on φ). Here are some possibilities:

a) $u(x,t) > 0$ everywhere in $\mathbb{R} \times (0,\infty)$;
b) $u(x,t)$ has compact support in x, but the fronts increase indefinitely: $\forall x_o \exists t_o > 0$: $u(x_o,t) > 0$ for $t > t_o$ ("there is no space localization");
c) $u(x,t)$ with compact support in x and $\exists L > 0$: $u(x,t) = 0$ for $|x| \geq L$ ("space localization").

These three possibilities do exist. But certainly more complicated behaviours for the fronts are possible; however, none is explicitly known.

We are convinced that the complete answer to the question "What happens when $P = \infty$?" must contain the description of all these phenomena. The following theorems are a first step in this direction.

THEOREM 1. *Suppose* $P = \infty$ *and*

$$\lim_{u \to 0} \varphi'(u)c'(u) > \frac{1}{2} . \tag{8}$$

If $u_o(x) \not\equiv 0$, *then* $u(x,t) > 0$ *for* $t > 0$, $x \in \mathbb{R}$.

REMARK. In Theorem 1 we do not consider the linear equations.

EXAMPLES.

a) For the equation (6), $m > 1$, the assumptions of Theorem 1 cannot be fulfilled: $P = \infty \Leftrightarrow n \geq 1$, (8) $\Leftrightarrow m + n \leq 2 \Rightarrow m \leq 1$.
b) Consider the equation ($u_o < 1$)

$$u_t = (u(\ell n \frac{1}{u})^{-\alpha})_{xx} - u(\ell n \frac{1}{u})^{\beta}, \quad \alpha \geq 0, \quad \beta \in \mathbb{R} . \tag{9}$$

Here $\varphi'(0) = 0$ if $\alpha > 0$, i.e. formally (9) is degenerated at $u = 0$. $P = \infty$ iff $\beta \leq 1$, $\lim \varphi'(u)c'(u) = \lim(\ell n \frac{1}{u})^{\beta-\alpha}$, i.e. (8) is fulfilled if $\beta \geq \alpha$. Thus, if $0 \leq \alpha \leq \beta \leq 1$, then $u(x,t) > 0$ ($t > 0$), i.e. the situation is the same as for the heat equation (1).

When $\beta > 1$, $u(x,t) = 0$ for $t \geq T_o$, $\forall\alpha$.
When $\alpha > 1$, the solution cannot be everywhere positive because the fronts exist in this case [7].

c) Consider a perturbation of the linear case weaker than (9):

$$u_t = (u(\ln \ln \frac{1}{u})^{-\alpha})_{xx} - u(\ln \ln \frac{1}{u})^{\beta}. \tag{10}$$

Here $P = \infty$ $\forall\beta \in \mathbb{R}$ and assumption (8) is equivalent to $\alpha \leq \beta$. We know from [7] that the result is the same for $\alpha > \beta$ too, thus in this sense the condition (8) is not necessary.

<u>THEOREM 2.</u> *Assume* $P = \infty$ *and*

$$\lim_{u \to 0} \varphi'(u)(\ln \int_u^M \frac{ds}{c(u)})^2 \leq \text{const}. \tag{11}$$

If $u_o(x_o) > 0$ *then* $u(x_o,t) > 0$ *for* $t > 0$ *arbitrary.*

<u>REMARK.</u> From $P = \infty$ and (11) follows that $\lim_{u \to 0} \varphi'(u) = 0$. The statement of Theorem 2 is obviously true if $\lim \varphi'(u) > 0$.

<u>EXAMPLES.</u>

a) For the equation (6) the assumptions of Theorem 2 are fulfilled if $n \geq 1$ - of course this is already known.

b) For the equation (9) the assumptions of Theorem 2 are fulfilled if $\alpha > 1$ and $\beta \geq 1$. In this case the fronts exist. It is known [8] that if $\alpha + \beta > 2$ then we have space localization, if $\alpha + \beta \leq 2$ we do not.

c) For equation (10), $P = \infty$ and (11) are equivalent to $\alpha \geq 2$, $\beta \in \mathbb{R}$.

d) The same conditions for

$$u_t = (u(\ln \ln \frac{1}{u})^{-\alpha})_{xx} - u(\ln \frac{1}{u})^{\beta}$$

are fulfilled if $\alpha \geq 2$ and $\beta \geq 1$.

REFERENCES

1. Antoncev, S.N.: On the localization of solutions of nonlinear degenerate elliptic and parabolic equations, Soviet Math. Dokl 24 (1981), 420-424.
2. Bénilan, P. and M.G. Crandall: The continuous dependence on φ of the solution of $u_t = \Delta\varphi(u)$, Indiana Univ. Math. J. 30 (1981), 161-177.
3. Diaz, G. and J.I. Diaz: Finite extinction time for a class of non-linear parabolic equations, Comm. Partial Differential Equations 4 (1979), 1213-1231.
4. Diaz, J.I. and L. Veron: Existence theory and qualitative properties of the solution of some first order quasilinear variational inequalities, Indiana Univ. Math. J. 32 (1983), 319-361.
5. Kalashnikov, A.S.: The propagation of disturbances in problems of nonlinear heat conduction with absorption. Z. Vycisl. Mat. i Mat. Fiz. 14 (1974), 891-905.
6. Kersner, R.: Degenerate parabolic equations with general nonlinearities, Nonlinear Anal. TMA 4 (1980), 1043-1061.
7. Kersner, R.: Filtration with absorption necessary and sufficient condition for the propagation of perturbation to have finite velocity, J. Math. Anal. Appl. 90 (1982), 463-4
8. Kersner, R.: Nonlinear heat conduction with absorption: space localization and extinction in finite time, SIAM J. Appl. Math. 43 (1983), 1274-1285.
9. Murray, F.D.: Perturbation effects on the decay of discontinuous solutions of nonlinear first order wave equations, SIAM J. Appl. Math., 19 (1970), 273-298.
10. Nagai, T.: Sufficient conditions for infinite propagation of solutions for semilinear heat equations, J. Math. Anal. Appl. (to appear).
11. Peletier, L.A.: The porous media equation. In "Applications of Nonlinear Analysis in the Physical Sciences", H. Amann, N. Bazley and K. Kirchgässner eds., pp. 229-241 (Pitman, 1981).

R. Kersner
Magyar Tudományos Akadémia
Számitástechnikai és Automatizálási
Kutató Intézete
P.O. Box 63
H-1502 Budapest
Hungary

H MATANO

Strong comparison principle in nonlinear parabolic equations

1. Introduction

In an earlier paper [5] the author has studied the dynamical structure of a class of equations in which a stronger version of the comparison principle holds. Such equations, characterized as "strongly order-preserving" local semiflows, include semilinear scalar parabolic equations in bounded spatial regions, weakly-coupled reaction-diffusion systems of the competition type with two unknowns, those of the cooperation type with any number of unknowns, and others. One of the interesting features of strongly order-preserving local semiflows having a certain compactness property is that any unstable equilibrium point has a nontrivial unstable set. This property, which is not trivial since the linearized instability is not assumed, has far more implications than are apparent on the dynamical structure of the semiflow. Another interesting property is that any positively-invariant bounded closed set, that is stable in the sense of Liapunov, contains at least one stable equilibrium point. As a corollary, one sees that between a super-solution above and a subsolution below, there always exists a stable equilibrium solution. An application of this useful theorem can be found in Matano and Mimura [8], in which it is shown that a competition-diffusion system can have a nonconstant stable equilibrium solution if the spatial region is something like a dumbbell-shaped one. This is a system version of the present author's former result on scalar equations [4]. The important point is that while the method in [4] heavily depends on the use of a nice Liapunov function, thereby being not applicable to general competiton-diffusion systems, the argument in [8] is simply based on the theory of S-O-P local semiflows and does not at all require such a Liapunov function to exist.

Hirsch [2], [3] has made an independent study of basically the same class of local semiflows (which he calls "strongly monotone", the definition being slightly more restrictive than our S-O-P; see below) and obtained other interesting results. Among other things, he has proved that almost all of the relatively compact orbits are quasi-convergent, that is, their ω limit sets consist only of equilibria. An important consequence of this theorem is that any periodic orbit is unstable.

The S-O-P local semiflows, though important from the applied point of view, of course form only a small class in the vast variety of dynamical systems. There are, however, some important equations that are not S-O-P in the strict sense of [5], [2], [3] but seem to come close to this category in unbounded space regions, some degenerate diffusion equations and some reaction-diffusion systems that are not necessarily weakly coupled. The last one is not even order-preserving. The original theory of [5], [2], [3] does not apply to these equations. Nonetheless, a careful study shows that results similar to those in [5], [2], [3] can be obtained for these equations. In particular, as for the latter two equations, a modified version of the original S-O-P theory can be constructed to cover them while preserving all the major theorems found in [2], [3], [5].

The aim of this paper is to present such a modified S-O-P theory and to show how the theory applies to specific problems such as the latter two of the above-mentioned problems. The first problem - i.e. equations in unbounded spatial regions - cannot be treated by this theory and requires a different approach; see [6], where a result similar to Theorem 3 below is presented.

In the next section we give main abstract results. Section 3 deals with applications of the theory. Because of limitations of space, detailed proof will not be given here. For further details, readers are referred to the author's forthcoming article [7].

2. Main results

Let X be an ordered metric space, with its metric and order relation denoted by d, $\geq$ respectively. By the term "ordered metric space", it is naturally required that the order structure of X and its topology be compatible, that is, $x_m \geq y_m$ $(m = 1,2,\ldots)$ and $x_m \to x_\infty$, $y_m \to y_\infty$ (as $m \to \infty$) imply $x_\infty \geq y_\infty$. We also assume that X is a complete metric space.

Let $\Phi = \{\Phi_t\}_{t\geq 0}$ be a local semiflow on X. More precisely, $\Phi : \mathrm{Dom}(\Phi) \to X$ is a continuous map, where $\mathrm{Dom}(\Phi)$ is an open subset of $X \times [0,\infty)$ satisfying the following conditions: if we set $\mathrm{Dom}(\Phi_t) \equiv \{x \in X \mid (x,t) \in \mathrm{Dom}(\Phi)\}$ for each $t \geq 0$ and define a map

$$\Phi_t : \mathrm{Dom}(\Phi_t) \to X$$

by $\Phi_t(x) = \Phi(x,t)$, then

(i) $\mathrm{Dom}(\Phi_o) = X$ and $\mathrm{Dom}(\Phi_t)$ is monotone nonincreasing in $t \geq 0$;

(ii) $\Phi_o(x) = x$ for every $x \in X$;

(iii) $\Phi_t \Phi_s = \Phi_{t+s}$ for every t, $s \geq 0$.

NOTATION 2.1. For each $x \in X$, we set

$$s(x) = \sup\{t \geq 0 \mid x \in \mathrm{Dom}(\Phi_t)\} \quad \text{(escape time)},$$

$$O^+(x) = \{\Phi_t(x) \mid 0 \leq t < s(x)\} \quad \text{(positive orbit)},$$

$$\omega(x) = \bigcap_{t \geq 0} \overline{O^+(\Phi_t(x))} \quad (\omega \text{ limit set}).$$

We also set

$$E = \{x \in X \mid \Phi_t(x) = x \text{ for all } t \geq 0\} \quad \text{(the set of equilibria)},$$

$$X^+(v) = \{z \in X \mid z \geq v\},\ X^-(v) = \{z \in X \mid z \leq v\},$$

$$[x,y] = \{z \in X \mid x \leq z \leq y\} \quad \text{(order interval)}.$$

DEFINITION 2.2. A local semiflow Φ on X is said to be order-preserving if $x \geq y$ implies $\Phi_t(x) \geq \Phi_t(y)$ for $t \in [0,s_o)$, where $s_o = \min\{s(x),s(y)\}$. Φ is called strongly order-preserving (S-O-P) if for any $x > y$ and any $t \in (0,s_o)$ there exist open sets $V \ni x$ and $W \ni y$ such that $\Phi_t(V) > \Phi_t(W)$.

DEFINITION 2.3. A positively invariant bounded closed set Y in X is said to be stable (in the sense of Liapunov) if for any open set $U \supset Y$ there exists an open set $V \supset Y$ such that $V \subset \mathrm{Dom}(\Phi_t)$ and $\Phi_t(V) \subset U$ for all $t \geq 0$. In particular, we define the stability of an equilibrium point in this way.

DEFINITION 2.4. Let X_1 be a positively invariant subset of X. Then one can define a local semiflow on X_1 by restricting the domain of Φ onto $\mathrm{Dom}(\Phi) \cap X_1 \times [0,\infty)$. We call this local semiflow the restriction of Φ onto X_1. An equilibrium point in X_1 is said to be X_1-stable if it is stable when Φ restricted onto X_1.

Our hypotheses are as follows:

(H.1) There exist subsets $X_o \subset X_1 \subset X$ such that

(a) X_o, X_1 are positively invariant and closed;

(b) for each $x \in X$ there exists a $T = T(x) \in [0,s(x))$ such that $\Phi_T(x) \in X_o$; moreover, $T : X \to \mathbb{R}^+$ is locally bounded;

(c) the restriction of Φ onto X_1 is S-O-P.

(H.2) Φ is <u>compact</u> in the sense that for any bounded set $B \subset X$ there exists a $\delta > 0$ such that $B \subset \mathrm{Dom}(\Phi_\delta)$ and that $\Phi_t(B)$ is relatively compact for each $t \in (0,\delta]$.

(H.3) For each $x \in X_o$, there exist sequences of points $\{y_m\} \subset X_1$, $\{z_m\} \subset X_1$ such that $y_m > x > z_m$ $(m = 1,2,\dots)$ and that $y_m \to x$, $z_m \to x$ as $m \to \infty$, where X_o, X_1 are as in (H.1).

<u>REMARK 2.5.</u> The original S-O-P theory in [2], [5] corresponds to the special case where $X_o = X_1 = X$, with X being a Banach space. In that case, (H.1), (H.3) simply reduce to the assumption that Φ is S-O-P.

<u>REMARK 2.6.</u> Hirsch's definition of strong comparison principle, which he calls "strongly monotone", is slightly more restrictive than our definition of S-O-P, since he requires that the order structure of X be defined by a closed cone having nonempty interior. His theory, therefore, does not apply to some important cases such as the case $X = L^p(\Omega)$ or the case $X = C_o(\overline{\Omega})$, that is, the space of continuous functions on $\overline{\Omega}$ vanishing on the boundary of Ω. Our definition, on the other hand, has no such restrictions, and, more important, most of his major theorems still hold true under the present hypotheses (see Theorems 4,5 and Lemma 2.12 below).

<u>REMARK 2.7.</u> Note that the hypotheses (H.1) ~ (H.3) do not even require Φ to be just order-preserving outside X_1. An example of such a local semiflow will be given in Section 3.

<u>THEOREM 1.</u> *Assume* (H.1), (H.2), *and let* $v \in E$. *Suppose* v *is not* X_1^+ *- stable, where* $X_1^+ = X_1 \cap X^+(v)$. *Then there exists a curve* $w : (-\infty,0] \to X_o$ *such that*

(i) $\Phi_t(w(s)) = w(t+s)$ *for any* $t \geq 0$, $s \leq 0$ *with* $t + s \leq 0$;

(ii) $w(t) > v$ *for any* $t \leq 0$;

(iii) $w(t) \to v$ *as* $t \to -\infty$.

If, instead, v *is not* X_1^-*-stable, then the above conclusion holds with* (ii) *being replaced by the condition $w(t) < v$.*

REMARK 2.8. Under the hypotheses (H.1), (H.2), one can easily show that an equilibrium point v is stable if and only if it is X_1-stable and that v is X_1-stable if and only if it is both X_1^+-stable and X_1^--stable. Theorem 1 therefore implies that any unstable equilibrium point has a nontrivial unstable set. Note that the linearized instability is not assumed here. This theorem is useful in the study of the dynamical structure of Φ.

THEOREM 2. *Let* X *be an ordered Banach space and assume* (H.1), (H.2). *Let* $v_1 < v_2$ *be a pair of equilibrium points such that* $[v_1,v_2] \cap E = \{v_1,v_2\}$ *and that* $[v_1,v_2]$ *is bounded. Then there exists an entire orbit connecting* v_1 *and* v_2.

REMARK 2.9. The above theorem follows from Theorem 1 and some degree of theoretical argument. The assumption that X be a Banach space can be replaced by a weaker assumption, namely that $[v_1,v_2]$ is homeomorphic to a retract of some Banach space.

THEOREM 3. *Assume* (H.1), (H.2), (H.3). *Let* Y *be a bounded, positively invariant, closed subset of* X. *Suppose* Y *is stable. Then it contains at least one stable equilibrium point.*

REMARK 2.10. Theorem 3 is exceedingly useful in finding stable equilibria. One of its important consequences is that between a strict supersolution above and a strict subsolution below there always exists a stable equilibrium solution. This type of theorem was first obtained by the author for semilinear diffusion equations [4; Theorem 4.2]. Weaker versions of Theorem 3 are found in [3], [5], and the present form of Theorem 3 in [7].

THEOREM 4. (Hirsch). *Assume* (H.1), (H.2), (H.3). *Then there exists no periodic orbit (apart from equilibria) that is orbitally stable.*

THEOREM 5. *Assume* (H.1), (H.2). *Assume further that* X_o *is bounded and that for any* $x \in X$ *there exists a* T^* *with* $T^* \geq T(x)$ *(where* $T(x)$ *is as in* (H.1)(b)) *such that*

$$\Phi_{T^*}(V) \cap (X_1^+(\Phi_{T^*}(x)) \cup X_1^-(\Phi_{T^*}(x)) \setminus \{\Phi_{T^*}(x)\}) \neq \emptyset$$

for any open neighbourhood V *of* x. *Then there exists an open dense subset* G *of* X *such that* $E \supset \omega(x) \neq \emptyset$ *for every* $x \in G$. *In other words, the orbit* $O^+(x)$ *stabilizes for generic* $x \in X$.

REMARK 2.11. Theorem 4 and a result analogous to Theorem 5 were obtained by Hirsch by using a limit set dichotomy theorem below (Lemma 2.12). Although his underlying hypothesis of strong monotonicity is stronger than ours, Lemma 2.12, and hence Theorems 4 and 5, still hold true under the present hypothesis of S-O-P. Note that Theorem 4 is also an immediate consequence of Theorem 3.

LEMMA 2.12. *Assume* (H.1), (H.2). *Let* $x > y$ *be points in* X_1 *such that both* $O^+(x)$ *and* $O^+(y)$ *are bounded. Then either* $\omega(x) > \omega(y)$ *or else* $\omega(x) = \omega(y) \subset E$.

Lastly in this section, we remark that the S-O-P hypothesis in (H.1)(c) can be replaced by a weaker condition, namely, what may be called "S-O-P with waiting time".

THEOREM 6. *Theorems 1 ~ 5 and Lemma 2.12 remains true if we replace* (H.1) (c) *by the following condition:*

(H.1)(c)' *the restriction of* Φ *onto* X_1 *is order-preserving; moreover, for any* $x, y \in X_1$ *with* $x > y$, *there exist* $t^* \geq 0$ *and open sets* $V \ni x$, $W \ni y$ *such that* $\Phi_{t^*}(V) > \Phi_{t^*}(W)$.

3. Applications

As mentioned in the Introduction, the S-O-P theory has proved to be a powerful method of qualitative analysis in many problems including competition-diffusion systems ([8]). In this section we give examples to which the original S-O-P theory of [2], [3], [5] does not apply unless modified as in the previous section.

EXAMPLE 3.1. Let Ω be a bounded domain in $\mathbb{R}^n$ with smooth boundary, and consider the following system:

$$\frac{\partial u}{\partial t} = d_1\Delta u + \vec{a}(x,u,v)\cdot\nabla u + f(x,u,v) \qquad (x \in \Omega,\ t > 0),$$
$$\frac{\partial v}{\partial t} = d_2\Delta v + \vec{b}(x,u,v)\cdot\nabla v + g(x,u,v) \qquad (x \in \Omega,\ t > 0), \tag{A}$$

where d_1, d_2 are positive constants and $\vec{a}$, $\vec{b}$, f, g are smooth functions satisfying $\partial f/\partial v < 0$, $\partial g/\partial u < 0$, so that (A) is a competition system. We impose either Neumann or Dirichlet boundary conditions. In the case $\partial\vec{a}/\partial v \equiv \partial\vec{b}/\partial u \equiv 0$, one can show that (A) defines a S-O-P local semiflow on $C(\overline{\Omega}) \times C(\overline{\Omega})$, with the order structure defined by $(\overline{u},\overline{v}) \geq (u,v) \Leftrightarrow \overline{u} \geq u, \overline{v} \leq v$ (here $C(\overline{\Omega})$ stands for $C_0(\overline{\Omega})$ in the case of the Dirichlet boundary conditions). The system (A), however, is in general not even order-preserving if $\partial\vec{a}/\partial v$ or $\partial\vec{b}/\partial u$ does not vanish identically.

To see how our modified S-O-P theory applies to the above problem, take, for simplicity, the specific case $f(u,v) = u(K_1 - \alpha u - \beta v)$, $g(u,v) = v(K_2 - \gamma u - \delta v)$ and consider nonnegative solutions of (A). Let $X = \{(u,v) \in C(\overline{\Omega}) \times C(\overline{\Omega}) \mid u \geq 0,\ v \geq 0\}$, $X_i = \{(u,v) \in C^1(\overline{\Omega}) \times C^1(\overline{\Omega}) \mid 0 \leq u \leq K_1 + r_i,\ 0 \leq v \leq K_2 + r_i,\ \|u\|_1 \leq M,\ \|v\|_1 \leq M\}$ $(i = 0,1)$, where $r_1 > r_0$ are arbitrary positive constants. Choosing $M > 0$ appropriately, one finds that the hypotheses (H.1), (H.2), (H.3) in the previous section are all satisfied provided that $\partial\vec{a}/\partial v$ and $\partial\vec{b}/\partial u$ are sufficiently small. Similar results hold if d_1, d_2 are functions of x, u, v.

EXAMPLE 3.2. Consider the degenerate diffusion equation

$$\frac{\partial u}{\partial t} = \frac{\partial^2}{\partial x^2}(u^m) + b(x,u)\frac{\partial u}{\partial x} + f(x,u) \qquad (x \in \mathbb{R},\ t > 0), \tag{B}$$

where $m > 1$. Let X be the space of continuous functions on $\mathbb{R}$ with compact support, with its topology defined by the uniform convergence on $\mathbb{R}$ together with the convergence of support. Under appropriate conditions on b, f, one can show that Theorem 6 in the previous section applies to (B) with $X_0 = X_1 = X$. For details, see [1].

REFERENCES

1. Crandall, M.G. and H. Matano (in preparation).
2. Hirsch, M.W. : Differential Equations and Convergence Almost Everywhere in Strongly Monotone Flows. Contemporary Mathematics 17, pp.267-285, (Amer. Math. Soc., 1983).

3. Hirsch, M.W.: The Dynamical Systems Approach to Differential Equations. Bull. Amer. Math. Soc. 11 (1984), 1-64.

4. Matano, H.: Asymptotic Behaviour and Stability of Solutions of Semilinear Diffusion Equations. Publ. Res. Inst. Math. Sci. Kyoto Univ. 15 (1979), 401-454.

5. Matano, H.: Existence of Nontrivial Unstable Sets for Equilibriums of Strongly Order-Preserving Systems. J. Fac. Sci. Univ. Tokyo 30 (1983), 645-673.

6. Matano, H.: L^∞ Stability of an Exponentially Decreasing Solution of the Problem $\Delta u + f(x,u) = 0$ in $\mathbb{R}^n$. Japan J. Appl. Math. 2 (1985), 85-110.

7. Matano, H.: Asymptotic Behaviour of Nonlinear Diffusion Equations. (Pitman, to appear).

8. Matano, H. and M. Mimura: Pattern Formation in Competition-Diffusion Systems in Nonconvex Domains. Publ. Res. Inst. Math. Sci. Kyoto Univ. 19 (1983), 1049-1079.

H. Matano
Department of Mathematics
Hiroshima University
NAKA-KU, Hiroshima 730
Japan

M MIMURA & T NAGAI

Asymptotic behaviour of the interface to a certain nonlinear diffusion-advection equation

1. Introduction

We consider the one-dimensional nonlinear degenerate diffusion and nonlocal advection equation

$$u_t = [(u^m)_x + (K*u)u]_x, \quad t > 0,\ x \in \mathbb{R} \tag{1.1}$$

with

$$u(x,0) = u_o(x), \quad x \in \mathbb{R}, \tag{1.2}$$

where $K*u = \int_{-\infty}^{\infty} K(x-\xi)u(\xi,t)d\xi$ with suitable kernel $K(x)$; $m \in (1,\infty)$ is a constant and $u_o(x)$ is a given nonnegative, bounded and integrable function in $\mathbb{R}$. From an ecological point of view, (1.1) is proposed as a caricature of Hamilton's model of selfish herd [2] to explain phenomenologically spatial aggregation of animal populations. Here $u(x,t)$ denotes the population density at point $x \in \mathbb{R}$ and at time $t > 0$. For the interpretation of the ecological background of (1.1), (1.2), see [3], [4].

In the absence of the advection term, (1.1) reduces to the well-known porous medium equation, which has been widely investigated by various authors. One of the most striking phenomena which this equation exhibits is the occurrence of <u>interfaces</u>. Suppose that $u_o(x) > 0$ on a bounded interval (a_1,a_2) and $u_o(x) \equiv 0$ on $\mathbb{R}\setminus(a_1,a_2)$. Then the interfaces between the region where $u > 0$ and the one where $u \equiv 0$ are described by the Lipschitz continuous curves $x = \rho_i(t)$ $(\rho_i(0) = a_i)$ $(i = 1,2)$, which are governed by the equation

$$\frac{d}{dt}\rho_i(t) = -\frac{m}{m-1}[u^{m-1}]_x(\rho_i(t),t) \quad (i = 1,2). \tag{1.3}$$

It is known that $\rho_1(t)$ (left interface) is either strictly decreasing with time $t > 0$ or <u>stationary</u> for a waiting time $t^* > 0$ and then strictly decreasing for $t > t^*$ (Aronson, Caffarelli and Vázquez [1], for instance). $\rho_2(t)$ has the similar properties in the opposite sense.

In contrast with the results above, there have been intensive studies on the behavior of interfaces to the porous medium equation with local advection terms such as a(u) in place of K*u in (1.1). This equation models the evaporation of a fluid through a porous medium. The results of Gilding and Diaz in this volume are relevant here. In this paper, we are interested in <u>nonlocal</u> aggregative advection terms of the type K*u, not only for ecological interest, but also as a prototype of a class of nonlinear degenerate diffusion equations with advection terms.

As a first step to study this sort of equation, we consider a simple kernel, namely $K(x) = k(2H(x)-1)$ with a constant $k \in (0, +\infty)$ where H(x) is the Heaviside step function. It then follows that

$$K*u = k\left(\int_{-\infty}^{x} u(\xi,t)d\xi - \int_{x}^{\infty} u(\xi,t)d\xi\right).$$

Taking advantage of the simplicity of the kernel, we are able to analyze almost completely the behavior of interfaces to (1.1), (1.2).

Let us assume the following conditions on the initial function u_o:

(A-1) $u_o \geq 0$ on $\mathbb{R}$, $u_o \in L^1(\mathbb{R}) \cap L^\infty(\mathbb{R})$;

(A-2) $(u_o^{m-1})_x \in L^\infty(\mathbb{R})$;

(A-3) $u_o > 0$ on a bounded interval (a_1,a_2) and
$u_o \equiv 0$ on $\mathbb{R} \setminus (a_1,a_2)$;

(A-4) u_o is piecewisely monotone on $\mathbb{R}$.

Then our main results are as follows:

(i)
$$u(x,t) \begin{cases} > 0, & x \in (\rho_1(t), \rho_2(t)), \\ \equiv 0 & \text{otherwise}, \end{cases} \qquad t > 0,$$
where $\rho_i(0) = a_i$ $(i = 1,2)$.

(ii) There exists a positive constant L such that

$$-L \leq \rho_1(t) < \rho_2(t) \leq L \quad \text{for all} \quad t \geq 0.$$

(iii) (The interface equation)

$$\frac{d}{dt}\rho_i(t) = -\frac{m}{m-1}(u^{m-1})_x(\rho_i(t),t) - (-1)^i k\int_{-\infty}^{\infty} u_o(x)dx \quad (1.4)$$

a.e. in $[0,\infty)$, $(i = 1,2)$.

(iv) $\lim_{t\to\infty} \rho_i(t) = \alpha_i$ with exponential order $(i = 1,2)$, where α_1 and α_2 are uniquely determined by

$$\alpha_1 = a_1 - \frac{\kappa}{2} + \frac{1}{c}\int_{a_1}^{a_2}\int_{x}^{a_2} u_o(y)dydx,$$

$$\alpha_2 = a_2 + \frac{\kappa}{2} - \frac{1}{c}\int_{a_1}^{a_2}\int_{a_1}^{x} u_o(y)dydx,$$

$$\kappa = k^{-1/m} c^{1-2/m}\int_0^1 [\xi(1-\xi)]^{-1/m}d\xi (=\alpha_2 - \alpha_1),$$

$$c = \int_{-\infty}^{\infty} u_o(x)dx.$$

(v) Let $t_i^* = \sup\{s:\rho_i(t) = a_i - (-1)^i kct\ (0 \leq t < s)\}$.
Then t_i^* is positive if and only if

$$\limsup_{x\to a_1+0} (x-a_1)^{-\frac{m+1}{m-1}}\int_{-\infty}^{x} u_o(\xi)d\xi < +\infty,$$

$$\limsup_{x\to a_2-0} (x_2-x)^{-\frac{m+1}{m-1}}\int_{x}^{+\infty} u_o(\xi)d\xi < +\infty.$$

The results (ii) - (v) contrast with the behavior of interfaces for the porous medium equation. In fact, (ii) implies that the spread of the solution is <u>localized</u> because of the aggregative effect. (iii) shows that the interfaces are not necessarily monotone. When $k = 0$, (1.4) reduces to (1.3). (iv) gives information on asymptotic behaviors not only for the solution $u(x,t)$, but also for the interfaces $\rho_i(t)$. Finally, (v) implies that the interface $\rho_1(t)$ (respectively $\rho_2(t)$) is increasing (respectively decreasing) with a <u>constant velocity</u> $kc = k\int_{-\infty}^{\infty} u_o(x)dx$ for $t \in (0,t_1^*)$ ($(0,t_2^*)$, respectively). Such a behavior corresponds to the existence of a positive waiting time for the interfaces when $k = 0$ (Vázquez [7]).

2. Existence and asymptotic behavior

The existence and uniqueness of solutions (in some generalized sense) to (1.1), (1.2) are shown in [5]. So we omit the detail. Here we only discuss the asymptotic behavior of the solution and of the interfaces for (1.1),(1.2).

DEFINITION. A stationary solution $\Omega(x)$ of the problem (1.1), (1.2) is any nonnegative function on $\mathbb{R}$ which satisfies $\Omega \in L^1(\mathbb{R}) \cap L^\infty(\mathbb{R})$ and

$$[(\Omega^m)_x + (K*\Omega)\Omega]_x = 0 \tag{2.1}$$

in the sense of distributions on $\mathbb{R}$.

THEOREM A. [4]. *There is a nontrivial stationary solution* $\Omega(x)$ *satisfying* $\int_{-\infty}^{\infty} \Omega(x)dx = c$, *for which there exist* $\alpha_1, \alpha_2 \in \mathbb{R}$ *such that the following properties hold:*

(i) $\Omega > 0$ *on* (α_1,α_2), $\Omega(x) \equiv 0$ *on* $\mathbb{R}\setminus(\alpha_1,\alpha_2)$;

(ii) $\Omega \in C^\infty((\alpha_1,\alpha_2))$;

(iii) $\int_{\alpha_1}^{\alpha_2}\int_{\alpha_1}^{x} \Omega(y)dydx = c\kappa/2$;

(iv) $$(\Omega^{m-1})_x(x) = \frac{k(m-1)}{m}\left(c - 2\int_{-\infty}^{x} \Omega(y)dy\right)$$ $$(\Omega^{m-1})_{xx} = -\frac{2k(m-1)}{m}\Omega(x)$$ *in* (α_1,α_2),

where $\kappa = \alpha_2 - \alpha_1 = k^{-1/m} c^{1-2/m} \int_0^1 [\eta(1-\eta)]^{-1/m}d\eta$.

THEOREM B [6]. *If, in addition to* (A.1), (A.2), u_0 *has compact support, then there are positive constants* C *and* λ *such that*

$$\|u(.,t) - \Omega\|_{L^\infty} \leq C\, e^{-\lambda t}$$

for sufficiently large time $t > 0$, *where* $\Omega(x)$ *is a stationary solution uniquely determined by*

$$\int_{-\infty}^{\infty} (\Omega(x) - u_o(x))dx = 0 \quad \textit{and} \quad \int_{-\infty}^{\infty}\int_{-\infty}^{x} (\Omega(y) - u_0(y))dydx = 0. \qquad (2.2)$$

Observe that, due to (2.2), the interface points of $\Omega(x)$, α_1 and α_2 are explicitly represented by

$$\alpha_1 = a_1 - \kappa/2 + \frac{1}{c}\int_{a_1}^{a_2}\int_{x}^{a_2} u_o(y)dydx$$

and

$$\alpha_2 = a_2 + \kappa/2 - \frac{1}{c}\int_{a_1}^{a_2}\int_{a_1}^{x} u_o(y)dydx,$$

respectively.

3. Interfaces

In the previous section we have found that the solution $u(x,t)$ approaches a stationary solution $\Omega(x)$, which is uniquely determined by the initial function $u_o(x)$, as t tends to $+\infty$. Here we are concerned with the asymptotic behavior of the interfaces $x = \rho_i(t)$ $(i = 1,2)$.

THEOREM C [6]. *Under* (A.1) - (A.4), *properties* (i) - (iv) *in section 1 hold.*

Finally, we show numerical simulations to supplement the assertion (v) in Section 1. Here we take $m = 2$, $k = 1$ and

$$u_o(x) = \begin{cases} \cos^n x, & |x| < \pi/2 \\ 0 & \text{otherwise;} \end{cases}$$

in this case $t_i^* > 0$ if and only if $n \geq 2$. Fig. 1 shows that the right interfaces for $n = 0.01, 0.3, 0.9, 1, 1.3$ and 1.7. It is observed that the interfaces are not straight. On the other hand, Fig. 2 shows the interfaces in the case when $n = 2, 3, 4, 5, 6, 7, 8, 9$ and 10. We clearly observe "straight" interface curves for $0 \leq t \leq t_2^*$. Details, including the proof of (v) in Section 1, will be given in a forthcoming paper.

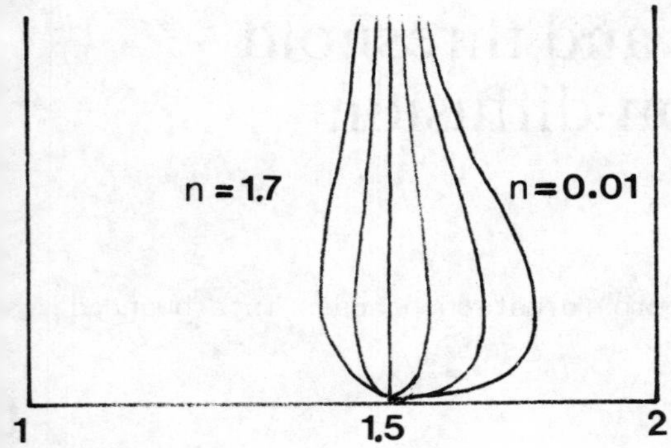

Fig. 1

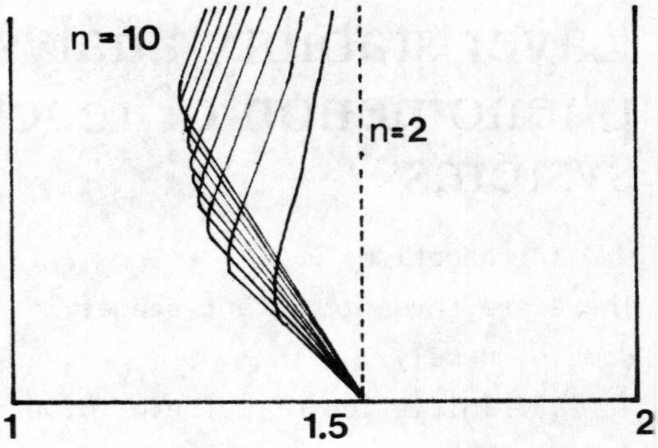

Fig. 2

REFERENCES

1. Aronson, D.G., L.A. Caffarelli and J.L. Vázquez: Interfaces with a corner point in one-dimensional porous medium flow (preprint).
2. Hamilton, W.D.: Geometry for the selfish herd, J. Theor. Biol. 31 (1971), 295-311.
3. Mimura, M.: Some convection-diffusion equations arising in population dynamics, Contemporary Mathematics 17, pp. 343-351 (Amer. Math. Soc., 1983).
4. Nagai, T. and M. Mimura: Asymptotic behavior for a nonlinear degenerate diffusion equation in population dynamics, SIAM J. Appl. Math. 43 (1983), 449-464.
5. Nagai, T. and M. Mimura: Some nonlinear degenerate diffusion equations related to population dynamics, J. Math. Soc. Japan 35 (1983), 539-562.
6. Nagai, T. and M. Mimura: Asymptotic behavior of the interfaces to a non-linear degenerate diffusion equation in population dynamics, submitted to Japan J. Appl. Math.
7. Vázquez, J.L.: The interfaces of one-dimensional flows in porous media, Trans. Amer. Math. Soc. 285 (1984), 717-737.

M. Mimura
Department of Mathematics
Hiroshima University
Naka-Ku, Hiroshima 730
Japan

T. Nagai
Department of Mathematics
Faculty of Education
Ehime University
Matsuyama 790
Japan

Y NISHIURA

Layer stability analysis and threshold phenomenon of reaction-diffusion systems

0. Introduction

There are three important stages in pattern formation process in a bounded domain, namely:

(1) Initiation of pattern formation,

(2) Transition state,

(3) Approach to the final state.

The instability of the basic state driven by some force (for instance, diffusion) becomes a trigger of the first step. This type of pattern formation is studied rather extensively by making use of bifurcation theory, though parameters are specialized and all patterns obtained are close to the basic state. Another type of initiation, which we are interested in, in this note, needs a big perturbation to the basic state which has to exceed some critical level called threshold. From the mathematical point of view, the second one is more difficult than the first one, since patterns appearing in the second case are, in general, of large amplitude and global dynamical problems should be treated. The second stage is more or less complicated and very difficult to describe thoroughly; however, we should note that a travelling front on $\mathbb{R}$ is one of the typical idealizations to describe the second state. The third one is related to problems of existence and stability for large amplitude steady states. Of course, these processes do not occur separately but occur successively.

In this note we will try to describe all of these processes for the following system ($0 < \varepsilon << 1$, $\sigma > 0$):

$$\begin{cases} u_t = \varepsilon^2 u_{xx} + f(u,v) & (t,x) \in (0,\infty) \times I,\ I = (0,1) \\ v_t = \frac{1}{\sigma} v_{xx} + g(u,v) & \\ u_x = 0 = v_x & \text{on } \partial I. \end{cases} \tag{P}$$

The smallness of the diffusion rate in the first equation enables us to apply the singular perturbation method to get the large amplitude steady states, which play a key role in the following. Although there are a number of examples in various models which show the above three steps successively, we focus on the following simple case. The functional forms of f and g are given in Fig. 1 (see the end of this introduction for the details). (P) has a constant state $\overline{U}$ which corresponds to the basic state of stage (1). If two nullclines intersect as in Fig. 1, then $\overline{U}$ is stable not only in ODE sense but also stable in PDE sense. Therefore, there are no bifurcations from $\overline{U}$, and in fact any small perturbation decays rapidly as in Fig. 4_a in Section 2. However, if one adds a large perturbation concentrated near the right boundary (see Fig. 4_b in Section 2)-namely, if the boundary trigger exceeds a critical level - it grows up and propagates like a travelling front; and finally settles down to the inner transition layer solution (ITL). We may call this <u>the threshold phenomenon at the boundary</u>. The natural question is "What is the threshold state or separatrix which consists of the

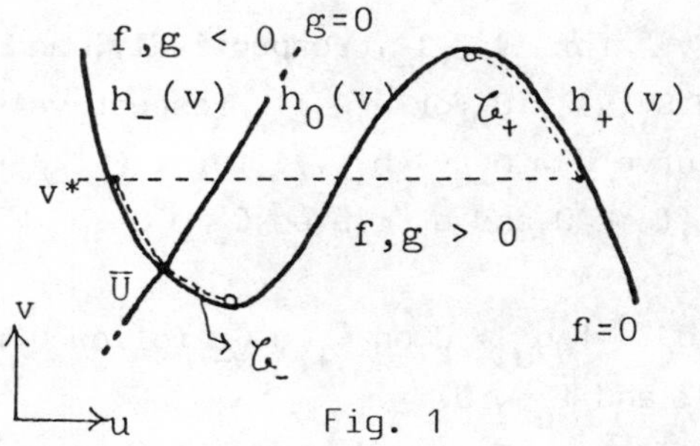

Fig. 1

boundary of two basins of $\overline{U}$ and ITL?" One can expect that this is given by the stable manifold of the boundary layer solution (BL) as in Fig. 5. In order to give a mathematical description of the above process, we have to solve three problems:

(A) Stability of ITL,

(B) Instability of BL and dim {unstable manifold} = 1,

(C) Connection problems - namely, the destinations of the unstable manifold of BL are the basic state $\overline{U}$ and the final state ITL, respectively.

ITL and BL can be constructed by using the singular perturbation method (see, for instance, [3], [2] and [4]). As for the stability problems of ITL and BL, they remain open so far (see [1]). The first stability result is obtained in [6] for small σ (see also [7]), and quite recently, they are solved systematically for a general σ in [7], [8] and [9]. In Section 1 we will outline the proof of the stability of ITL, and the instability of BL will be made clear by comparing the behaviors of the principal eigenvalues of the associated Sturm-Liouville problems. Unfortunately, the problem (C) is still open in a general setting; however, we may prove it rigorously for the limiting case $\sigma \downarrow 0$ with special nonlinearities. We discuss this point briefly in Section 2. We assume the following for f and g.

(A.1) The nullcline of f is sigmoidal and consists of three curves $u = h_-(v)$, $h_0(v)$ and $h_+(v)$, with $h_-(v) < h_0(v) < h_+(v)$.

(A.2) $J(v)$ has an isolated zero at $v = v^*$, such that $dJ/dv < 0$ at

$$v = v^*, \text{ where } J(v) = \int_{h_-(v)}^{h_+(v)} f(s,v)ds.$$

(A.3) Let $G_\pm(v) = g(h_\pm(v),v)$ for $v \in I_\pm$, respectively, and $v^* \in I_+ \cap I_-$. Then, $dG_\pm/dv < 0$ and $G_\pm(v) \gtrless 0$ for $v \in I_\pm$, respectively.

(A.4) Let C_+ (C_-) be the curve $u = h_+(v)(h_-(v))$ for $v \geq (\leq) v^*$, respectively (see Fig. 1). Then, $f_u < 0$ and $g_v < 0$ on $C_+ \cup C_-$.

<u>REMARK 0.1.</u> Note that $f_u g_v - f_v g_u > 0$ on $C_+ \cup C_-$ follows from $f(h_\pm(v),v) = 0$, $dG_\pm/dv < 0$, and $f_u < 0$.

1. Stability and instability of layer solutions

First we present two existence theorems. Theorem 1 shows the existence of singularly perturbed solutions which connect the h_--state to the h_+-state by an inner transition layer. Theorem 2 exhibits boundary layer solutions which are close to the basic state $\overline{U}$ except near the boundary. We fix σ to be an appropriate value such that the following two theorems hold, since we do not discuss the σ-dependence of layer solutions in this note. We denote by (SP) the stationary problem of (P).

<u>THEOREM 1.</u> ([3], [2]) *There exists an* $\varepsilon_0 > 0$ *such that* (SP) *has an* ε*-family of solutions of* ITL $U^\varepsilon = (u(x;\varepsilon),v(x;\varepsilon))$ *for* $0 < \varepsilon < \varepsilon_0$. U^ε *is uniformly bounded in* $C^2_\varepsilon \times C^2$ *and satisfies*

$$\lim_{\varepsilon\downarrow 0} u(x;\varepsilon) = U^*(x) \overset{\text{def}}{=} \begin{cases} h_-(V^*(x)), & x \in [0,x^*) \\ h_+(V^*(x)), & x \in (x^*,1], \end{cases}$$

uniformly on $I \backslash I_\kappa$ *for any* $\kappa > 0$ *(see Fig. 2), and*

$$\lim_{\varepsilon\downarrow 0} v(x;\varepsilon) = V^*(x) \quad \textit{uniformly},$$

where $V^*(x)$ *is a monotone increasing solution of* $\frac{1}{\sigma} V_{xx} + G(V) = 0$ *in* I *with* $V_x = 0$ *on* ∂I *and* $G(V) = G_+(V)(G_-(V))$ *for* $V \geq (\leq) v^*$. *Here* x^* *indicates the layer position, uniquely determined by* $V^*(x) = v^*$, *and* $I_\kappa = (x^*-\kappa, x^*+\kappa)$.

<u>THEOREM 2.</u> ([4], [9]) *There exists an* $\varepsilon_1 > 0$ *such that* (SP) *has an* ε*-family of solutions of* BL, $U_B^\varepsilon = (u_b(x;\varepsilon), v_b(x;\varepsilon))$ *for* $0 < \varepsilon < \varepsilon_1$. U_B^ε *is uniformly bounded in* $C_\varepsilon^2 \times C^2$, *and satisfies (see Fig. 3)*

$$\lim_{\varepsilon\downarrow 0} u_b(x;\varepsilon) = \overline{u} \qquad \textit{compact uniformly in } [0,1),$$

$$\lim_{\varepsilon\downarrow 0} v_b(x;\varepsilon) = \overline{v} \qquad \textit{uniformly on } I,$$

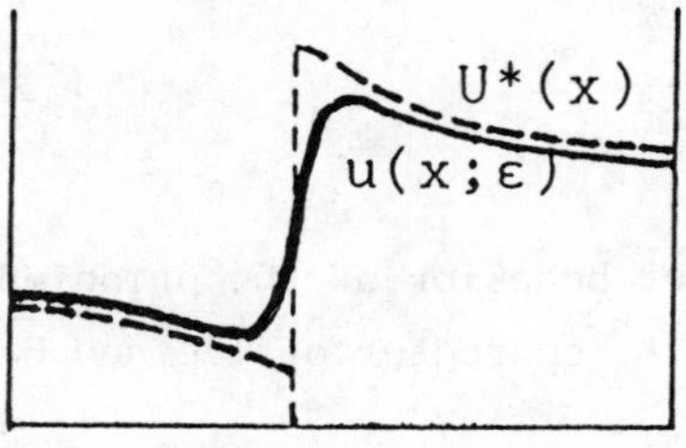

Fig. 2 u-component of ITL

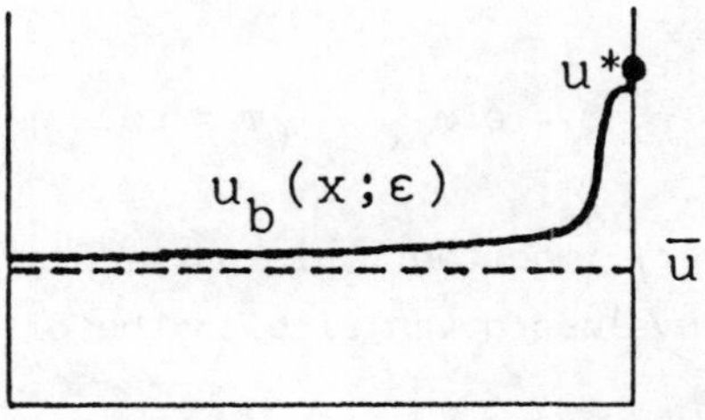

Fig. 3 u-component of BL

where $\overline{u}(\overline{v})$ *is the* $u(v)$*-component of the basic state* $\overline{U}$. *The depth of the boundary layer* $u^* = \lim_{\varepsilon\downarrow 0} u_b(1,\varepsilon)$ *is determined by* $\int_{\overline{u}}^{u^*} f(s,\overline{v})\, ds = 0$.

The stability properties of ITL and BL are determined by the spectral distribution of the following linearized eigenvalue problem

$$\begin{aligned} &\varepsilon^2 w_{xx} + f_u^\varepsilon w + f_v^\varepsilon z = \lambda w, \\ & \qquad\qquad\qquad\qquad\qquad\qquad \text{in} \quad I, \\ &\frac{1}{\sigma} z_{xx} + g_u^\varepsilon w + g_v^\varepsilon z = \lambda z, \qquad\qquad\qquad\qquad (LP) \\ &w_x = 0 = z_x \qquad\qquad\qquad \text{on} \quad \partial I, \end{aligned}$$

where all partial derivatives are evaluated at ITL or BL. It is convenient to divide the spectrum of (LP) into two classes; one consists of noncritical eigenvalues which are bounded away from zero for small ε, the other of critical eigenvalues which go to zero as $\varepsilon \downarrow 0$. Then we can prove the following results.

STABILITY THEOREM OF ITL ([7], [8]). *There exists only one critical eigenvalue* $\lambda = \lambda_I(\varepsilon)$ *of* (LP) *at* ITL, *which is real and simple. When* $\varepsilon \downarrow 0$, *it behaves as* $\lambda_I(\varepsilon) \simeq -\gamma\varepsilon(\gamma > 0)$. *The remaining noncritical eigenvalues have strictly negative real parts for small* ε.

INSTABILITY THEOREM OF BL ([9]). (LP) *at* BL *has only one real, positive, simple eigenvalue* $\lambda = \lambda_B(\varepsilon)$ *with* $\lim_{\varepsilon\downarrow 0} \lambda_B(\varepsilon) = \lambda_B^* > 0$. *All the other eigenvalues have strictly negative parts for small* ε.

In order to solve (LP), the spectral behavior of the Sturm-Liouville problem

$$L^\varepsilon\varphi = \varepsilon^2\varphi_{xx} + f_u^\varepsilon\varphi = \zeta\varphi, \quad \varphi_x = 0 \quad \text{on} \quad \partial I$$

is very important. In particular, the different behavior of the principal eigenvalues gives rise to the different stability character of ITL and BL.

LEMMA 1.1. *Let* $\zeta_I^0(\varepsilon)$ *be the principal eigenvalue of* L^ε *at* ITL. *Then,* $\zeta_I^0(\varepsilon)$ *is positive and tends to zero when* $\varepsilon \downarrow 0$ *as* $\zeta_I^0 = \varepsilon\hat{\zeta}_I(\varepsilon)$, *where* $\hat{\zeta}_I(\varepsilon)$ *is continuous and* $\zeta_I^* = \hat{\zeta}_I(0) > 0$. *On the other hand, the principal eigenvalue* $\zeta_B^0(\varepsilon)$ *in the* BL *case is strictly positive for small* ε. *All the other eigenvalues are strictly negative in both cases.*

In the following we only consider the ITL case.

Noting the solubility condition, the first equation of (LP) can be solved to give $w = (L^\varepsilon - \lambda)^{-1}(-f_v^\varepsilon z)$. Substitution into the second equation of (LP) leads to

$$\frac{1}{\sigma} z_{xx} + \frac{\langle -f_v^\varepsilon z,\, \varphi_o^\varepsilon\rangle}{\zeta_I^0 - \lambda} g_u^\varepsilon\varphi_o^\varepsilon + g_u^\varepsilon(L^\varepsilon-\lambda)^\dagger(-f_v^\varepsilon z) + g_v^\varepsilon z = \lambda z, \tag{1}$$

where $\langle , \rangle$ denotes the L^2-inner product and $(L^\varepsilon-\lambda)^\dagger(u) = \sum_{n\geq 1}\langle u,\varphi_n^\varepsilon\rangle\varphi_n^\varepsilon/(\zeta_n^\varepsilon-\lambda)$.

Here $\{\zeta_n^\varepsilon, \varphi_n^\varepsilon\}$ is the complete orthonormal set of L^ε at ITL. We focus on the asymptotic behavior of the critical eigenvalues, since the noncritical ones are not dangerous for the stability of ITL (see [8] for details). Whether or not we can solve (1) depends uniquely on the characterization of the second and the third terms of (1) as $\varepsilon \downarrow 0$. The following two lemmas answer this question.

LEMMA 1.2. $(L^\varepsilon-\lambda)^\dagger(-f_v^\varepsilon\cdot)$ *becomes a multiplication operator when* $\varepsilon \downarrow 0$, *namely*

$$(L^\varepsilon-\lambda)^\dagger(-f_v^\varepsilon u) \xrightarrow[\varepsilon \downarrow 0]{} \frac{-f_v^*}{f_u^*-\lambda} u$$

in L^2*-strong sense for any bounded* $u \in L^2(I)$ *and* $\mathrm{Re}\lambda > \max_{x \in I} f_u^*$. *Here* $f_u^* = f_u(U^*(x), V^*(x))$ *and so on; the above convergence is uniform on any bounded set in* $H^1(I)$.

LEMMA 1.3.

$$\frac{-f_v^\varepsilon \varphi_o^\varepsilon}{\sqrt{\varepsilon}} \left(\frac{g_u^\varepsilon \varphi_o^\varepsilon}{\sqrt{\varepsilon}}\right) \xrightarrow[\varepsilon \downarrow 0]{} c_1^*\delta^* \ (c_2^*\delta^*)$$

in H^{-1}*-sense, where* δ^* *is a Dirac* δ*-function at* $x = x^*$, *i.e.,* $\delta^* = \delta(x-x^*)$, *and* $c_i^*(i = 1,2)$ *are uniquely determined positive constants.*

Using these lemmas, we can derive a limiting problem of (1) as $\varepsilon \downarrow 0$ in the following weak form

$$\frac{1}{\sigma}\langle z_x, \psi_x\rangle - \frac{c_1^* c_2^*}{\zeta_1^* - \tau_o}\langle z, \delta^*\rangle\langle \delta^*, \psi\rangle - \langle \frac{\det^*}{f_u^*} z, \psi\rangle = 0, \ z \in H_N^1(I), \qquad (2)$$

for any $\psi \in H^1(I)$, where $\det^* = f_u^* g_v^* - f_v^* g_u^*$ which is positive from Remark 0.1, and $H_N^1(I)$ denotes the closure of $\{\cos(n\pi x)\}_{n \geq 0}$ in $H^1(I)$-space. Here we assume that the critical eigenvalue λ is of the from $\lambda = \varepsilon\tau(\varepsilon)$ with $\tau_o = \tau(0)$. For the justification of this hypothesis, see [8] or [7]. We call (2) the singular limit eigenvalue problem, which reserves the information from layer part as the Dirac δ-function. What we need is to know the sign of τ_o in (2). Note that U_x^ε is an eigenfunction of (LP) under Dirichlet boundary conditions associated with the zero eigenvalue. Comparing this limit as $\varepsilon \downarrow 0$ with the solution of (2) under Neumann boundary conditions, we obtain $\tau_0 < 0$, which implies the Stability Theorem of ITL.

2. Threshold phenomenon and connection problem

In view of the numerical simulations in Fig. 4, we can see that the final state depends on the size of the deviation from the basic state $\overline{U}$ at the boundary. Small perturbations are not effective in that they decay to $\overline{U}$ promptly. However, if the perturbation is large enough, a front is formed near the boundary, then it moves to the left like a travelling wave with adjusting the values of the outer parts to those of ITL, and eventually settles down to ITL. We may call this the threshold phenomenon at the boundary. Note also that not only the size, but also the position of the perturbation is important in the above pattern formation. Thus, Fig. 4

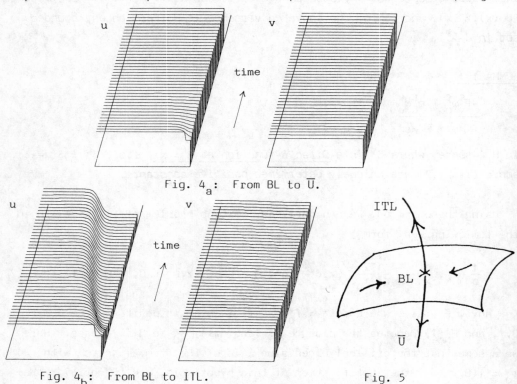

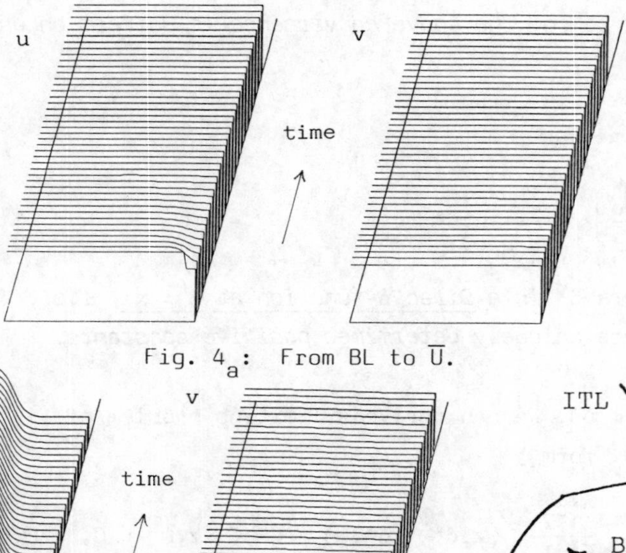

Fig. 4_a: From BL to $\overline{U}$.

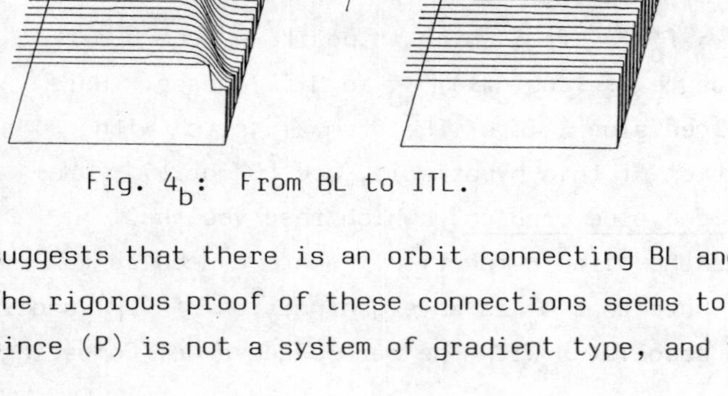

Fig. 4_b: From BL to ITL.

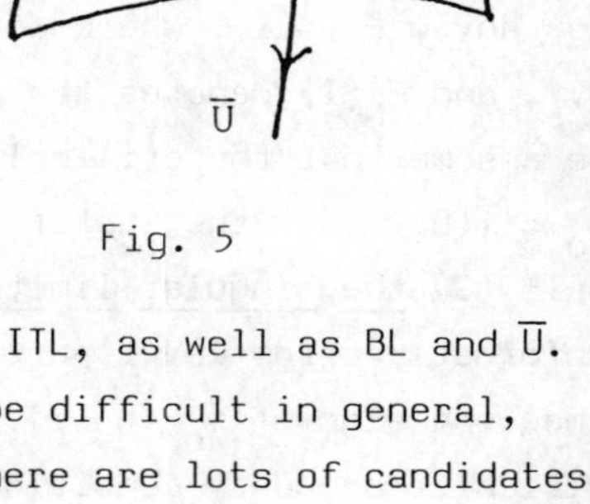

Fig. 5

suggests that there is an orbit connecting BL and ITL, as well as BL and $\overline{U}$. The rigorous proof of these connections seems to be difficult in general, since (P) is not a system of gradient type, and there are lots of candidates for the destinations of the unstable manifold (i.e., coexistence of multiple stable states). However, there is a possibility to show it rigorously, if one goes to the limiting system of (P) as $\sigma \downarrow 0$, i.e.

$$\begin{aligned} &u_t = \varepsilon^2 u_{xx} + f(u,\xi), \\ &\xi_t = \int_I g(u,\xi)\, dx, \\ &u_x = 0 \qquad \text{on } \partial I, \end{aligned} \tag{$(P)_o$}$$

where $v \equiv \xi$ is a constant function (see [5] and [10]). Moreover, it is known that $(P)_o$ has a Lyapounov function (see [10]) when the nonlinearities take the following form

$$f(u,v) = f_o(u) - v, \quad g(u,v) = c(du - v), \ (c > d). \tag{3}$$

Consequently, if one considers the connection problem for $(P)_o$ with (3), one can see that the destinations of the unstable manifold must be steady states, $(P)_o$ preserves the monotonicity of u, and the monotone increasing (or decreasing) ITL is unique. These observations suggest a possibility of analytical proof for the connection problem, which will be discussed somewhere.

REFERENCES

1. Conway, E.D.: Research Notes in Math., Pitman, 101 (1984), 85-133.
2. Ito, M.: Hiroshima Math. J., 14(1985), 619-629.
3. Mimura, M., M. Tabata and Y. Hosono.: SIAM J. Math. Anal., 11 (1980), 613-631.
4. Mimura, M., Y. Nishiura, A. Tesei and T. Tsujikawa.: Hiroshima Math. J., 14 (1984), 425-449.
5. Nishiura, Y.: SIAM J. Math. Anal., 13 (1982), 555-593.
6. Nishiura, Y. and H. Fujii: An approach to the stability of singularly perturbed solutions in reaction-diffusion systems (preprint).
7. Nishiura, Y. and H. Fujii: Stability theorem for singularly perturbed solutions to systems of reaction-diffusion equations, to appear in Proc. of Japan Acad.
8. Nishiura, Y. and H. Fujii: in preparation.
9. Nishiura, Y.: in preparation.

10. Nishiura, Y. and Gardner, R: in preparation.

Y. Nishiura
Institute of Computer Sciences
Kyoto Sangyo University
Kamigamo-Motoyama, Kita-Ku
Kyoto 603
Japan

L A PELETIER & A TESEI

The critical size problem in a spatially varying habitat

1. Introduction

Suppose a population lives in a habitat Ω in which there exists a region Ω^+ where the population flourishes and multiplies, while eslewhere, in the region Ω^- outside Ω^+ the population perishes. Then, depending on the relative sizes of Ω^+ and Ω^-, the dispersal and growth rates of the population, and the conditions at the boundary $\partial\Omega$ of Ω, the population may die out, tend to a stable distribution or grow explosively. The problem of determining which of these three possibilities will occur is a classical one in population dynamics; for references we refer to Okubo [8].

In this paper we shall discuss this question for a single species population on the basis of a model due to Gurney and Nisbet [4,5] and Gurtin and MacCamy [6]. According to this model, the population density u(x,t) satisfies the diffusion equation

$$u_t = D\Delta(u^m) + f(x,u) \qquad x \in \Omega,\ t > 0 \qquad (1.1)$$

in which $D > 0$ and $m > 1$ to model a tendency to avoid crowding. Following Gurney and Nisbet [4,5] we set

$$f(x,u) = g(x)u, \qquad (1.2)$$

where g(x) is the local growth rate; it has the properties

$$g(x) > 0 \quad (< 0) \quad \text{in} \quad \Omega^+ \quad (\Omega^-).$$

As to the situation at the boundary $\partial\Omega$ of the habitat, we distinguish two cases.

Case D. The individuals are immediately removed:

$$u(x,t) = 0 \qquad x \in \partial\Omega, \quad t > 0. \qquad (1.3a)$$

(A Dirichlet boundary condition).

Case N. The individuals are "fenced in", i.e. no individuals can enter or leave the habitat:

$$(u^m)_\nu(x,t) = 0 \qquad x \in \partial\Omega, \quad t > 0, \tag{1.3b}$$

where the subscript ν denotes differentiation in the direction of the outward normal ν on $\partial\Omega$. (A Neumann boundary condition).

We shall assume that D, m and g(x) are given quantities and we wish to determine the effect of the size of Ω on the existence of a stable population distribution in the Cases D and N.

In an earlier paper, Namba [7] discussed this question - partly analytically and partly numerically - when $\Omega = (-L,L)$ and g satisfies the hypotheses:

H1. $g \in C^1(\mathbb{R})$;

H2. $g(-x) = g(x)$ for all $x \in \mathbb{R}$;

H3. $xg'(x) < 0$ for all $x \in \mathbb{R}\setminus\{0\}$

H4. $g(\pm L_o) = 0$ for some $L_o > 0$;

and in particular, when $g(x) = A - Bx^2$ (Fig. 1).

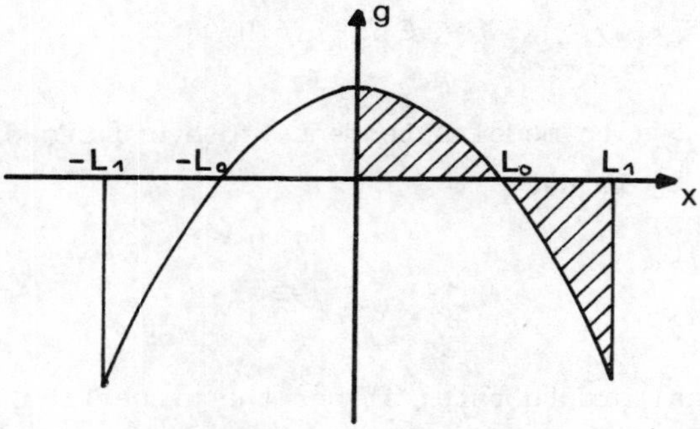

Fig. 1. The growth rate g(x).

Namba found for Case N what appeared to be globally stable stationary solutions provided $L > L_1$, where L_1 is (uniquely) determined by the condition

$$\int_0^{L_1} g(x)dx = 0, \tag{1.4}$$

and he proved that there could be no stationary solutions when $L < L_1$. Thus, the critical minimum size for $\Omega = (-L,L)$ to support a stable population would seem to be $2L_1$.

In this paper we shall prove this conjecture, show that for Case D the critical minimum size is zero, and establish the uniqueness and global stability of the stationary solution when it exists.

2. Stationary solutions

Suppose v is a stationary solution of (1.1). Then due to the symmetry of g, v is symmetric with respect to x = 0 and satisfies

$$\begin{cases} (v^m)'' + g(x)v = 0 \\ v(0) = c,\ v'(0) = 0 \end{cases} \tag{I}$$

where c is some positive number. For convenience we have absorbed the diffusion coefficient D in the function g. For every c > 0 this problem has a unique symmetric solution v(x,c) in a neighbourhood of x = 0, which we can continue to larger values of $|x|$ as long as v > 0.

Set

$$\xi(c) = \sup\{x > 0 : v(.,c) > 0 \text{ and } v'(.,c) < 0 \text{ on } (0,x)\}.$$

Since $g(0) > 0$, $\xi(c)$ is well defined and positive for every c > 0. In the following proposition we list a number of properties of ξ.

PROPOSITION. *Suppose* g *satisfies* H1-4. *Then*

(i) $\xi \in C(0,\infty)$, $\xi > 0$ *on* $(0,\infty)$.

There exists a number $c^* > 0$ *such that*

(ii) $\xi(c)$ *is strictly increasing on* $(0,c^*)$

$\xi(c)$ *is strictly decreasing on* (c^*,∞);

(iii) $\lim_{c\to 0} \xi(c) = 0$ *and* $\lim_{c\to\infty} \xi(c) = L_1$.

(iv) $0 < c \le c^* \Rightarrow v(\xi(c),c) = 0$

$c^* \le c < \infty \Rightarrow v'(\xi(c),c) = 0$.

Thus the graph of ξ consists of two monotone branches, the Dirichlet branch Γ_D ($c < c^*$) and the Neumann branch Γ_N ($c > c^*$), which join at $c = c^*$ (see Fig. 2). For the proof of this proposition we refer to [9].

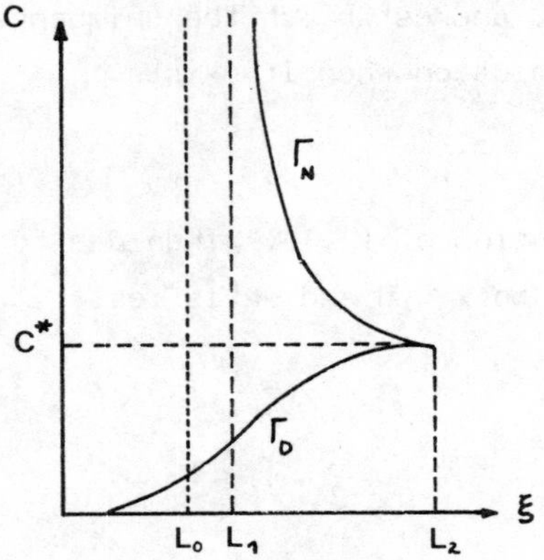

Fig. 2. The graph of $\xi(c)$.

The existence and uniqueness of stationary solutions in the Cases D and N can be deduced from the properties of the function c. (See Fig. 2). Set

$$\xi(c^*) = L_2.$$

(i) If $0 < L \leq L_1$, there exists a unique number $c_D(L)$ such that $(L, c_D(L)) \in \Gamma_D$ and no number $c > 0$ such that $(L,c) \in \Gamma_N$. Thus there exists precisely one stationary solution in Case D and none in Case N.

(ii) If $L_1 < L < L_2$, there exists one number $c_D(L)$ such that $(L, c_D) \in \Gamma_D$ and one number $c_N(L)$ such that $(L, c_N) \in \Gamma_N$. Therefore in each case there exists a unique stationary solution.

(iii) If $L \geq L_2$ we define the function

$$v_L(x) = \begin{cases} v(x, c^*) & |x| < L_2 \\ 0 & L_2 \leq |x| \leq L. \end{cases}$$

It is readily seen that v_L is a stationary solution for both cases. We summarize these results in the following theorem.

<u>THEOREM 1.</u> *Suppose the growth rate* $g(x)$ *satisfies* H1-4.

Case D: *There exists for every* $L > 0$ *a unique stationary solution* $v_{D,L}$ *of*

(1.1) - (1.3).

Case N: *There exists for every* $L > L_1$ *a unique stationary solution* $v_{N,L}$ *of* (1.1) - (1.3) *and no stationary solution if* $0 < L < L_1$.

If $L_1 < L < L_2$, *then*

$$v_{D,L}(x) < v_{N,L}(x) \quad \textit{for all} \quad x \in [-L,L]$$

and if $L > L_2$, *then*

$$v_{D,L}(x) = v_{N,L}(x) \quad \textit{for all} \quad x \in [-L,L].$$

The details of the proof can be found in [9].

3. Stability

The evolution of the population density u with time is described by the problem

$$\begin{cases} u_t = (u^m)_{xx} + g(x)u & \text{in } (-L,L) \times \mathbb{R}^+ \\ u = 0 \text{(Case D)} \quad \text{or} \quad (u^m)_x = 0 \text{(Case N)} & \text{on } \{-L,L\} \times \mathbb{R}^+ \\ u(x,0) = u_o(x) & \text{in } [-L,L], \end{cases} \tag{II}$$

where, as before, the diffusion coefficient D has been absorbed in the function g and the time t. Here u_o is an arbitrary given initial density; naturally $u_o \geq 0$ and $u_o \not\equiv 0$. We shall assume that u_o is continuous. By a solution u of Problem II we shall mean a weak solution in the sense defined in [2] and [9]. Emphasizing its dependence on u_o we shall write it as $u(t,u_o)$.

For convenience we shall write $\Omega_L = (-L,L)$; recall that the "sanctuary" is given by

$$\Omega^+ = \{x \in \mathbb{R} : g(x) > 0\} = (-L_o,L_o).$$

Suppose

$$\text{supp } u_o \cap \Omega^+ \neq \emptyset. \tag{3.1}$$

Then there exists a point $x_o \in (-L,L)$ such that

$$u_o(x_o) > 0 \quad \text{and} \quad g(x_o) > 0.$$

Now consider the problem

$$\begin{cases} (w^m)'' + g(x)w = 0 & \text{in } I = (x_o-\delta,\ x_o+\delta) \quad (\delta > 0) \\ w(x_o-\delta) = w(x_o+\delta) = 0. \end{cases}$$

For δ sufficiently small it has a unique positive solution w_δ such that $u_o \geq w_\delta$ in I_δ. The function

$$\underline{u}(x) = \begin{cases} w_\delta(x) & x \in I_\delta \\ 0 & x \in \Omega_L \setminus I_\delta \end{cases}$$

is a subsolution of Problem II for both Cases D and N.

To construct a supersolution, we consider for $c > c^*$ the family of functions

$$z(x,c) = \begin{cases} v(x,c) & \text{if } |x| < \xi(c) \\ v(\xi(c),c) & \text{if } |x| \geq \xi(c). \end{cases}$$

For Case D they are supersolutions of Problem II, and if we choose c large enough, $z(.,c) \geq u_o$ in Ω_L. For Case N they are supersolutions provided $L \geq \xi(c)$. As long as $L > L_1$ this can be achieved by choosing c large enough. Again we can thereby ensure that $z(.,c) \geq u_o$.

Thus we have constructed a subsolution $\underline{u}$ for any $L > 0$ and a supersolution, which we denote by $\overline{u}$, for any $L > 0$ in Case D and any $L > L_1$ in Case N. Clearly $\underline{u} < \overline{u}$ in Ω_L.

It is well-known that

$$u(t,\underline{u}) \nearrow v^- \quad \text{and} \quad u(t,\overline{u}) \searrow v^+ \quad \text{as} \quad t \to \infty,$$

where v^- and v^+ are stationary solutions of Problem II. By Theorem 1 they are unique, whence $v^- = v^+ := v$. By construction $\underline{u} \leq u_o \leq \overline{u}$, and hence

$$u(t,\underline{u}) \leq u(t,u_o) \leq u(t,\overline{u}) \quad \text{for all} \quad t \geq 0,$$

which means that

$$u(t,u_o) \to v \quad \text{as} \quad t \to \infty.$$

It remains to consider Case N, when $0 < L \leq L_1$. Then there exists no stationary solution, and $u(t,u_o)$ becomes unbounded as $t \to \infty$. Indeed, if $u(t,u_o)$ would be bounded, it would converge as $t \to \infty$ and the limit would be an equilibrium solution, which we know does not exist.

We are now ready to formulate our stability theorem.

<u>THEOREM 2.</u> *Suppose the growth function* $g(x)$ *satisfies* H1-4 *and*

$$\text{supp } u_o \cap \Omega^+ \neq \emptyset.$$

Then, if $L > 0$ (Case D) *or* $L > L_1$ (Case N),

$$u(t,u_o) \to v \quad \text{as} \quad t \to \infty,$$

where v *is the unique stationary solution of Problem II. If* $0 < L \leq L_1$ *in* Case N, *then* $u(t,u_o)$ *grows without bound as* $t \to \infty$.

If

$$\text{supp } u_o \cap \Omega^+ = \emptyset$$

it may still happen that at some later time t_o the population reaches the sanctuary Ω^+:

$$\text{supp } u(t_o,u_o) \cap \Omega^+ \neq \emptyset.$$

In that case, according to Theorem 2, however small the foothold, the population will grow and eventually approach the steady state if it exists.

On the other hand, the population may also never reach the sanctuary and eventually die out [3].

REFERENCES

1. Aronson, D.G.: Bifurcation phenomena associated with nonlinear diffusion mechanisms. In "Partial Differential Equations and Dynamical Systems", W.E. Fitzgibbon III Ed. (Pitman, 1984).
2. Aronson, D.G., M.G. Crandall and L.A. Peletier: Stabilization of solutions of a degenerate nonlinear diffusion problem, Nonlinear Anal. TMA 6 (1982), 1001-1022.
3. Bertsch, M., T. Nanbu and L.A. Peletier: Decay of solutions of a degenerate nonlinear diffusion equation, Nonlinear Anal. TMA 6 (1982), 539-554.
4. Gurney, W.S.C. and R.M. Nisbet: The regulation of inhomogeneous populations, J. Theor. Biol. 52 (1975), 441-457.
5. Gurney, W.S.C. and R.M. Nisbet: A note on nonlinear population transport, J. Theor. Biol. 56 (1976), 249-251.
6. Gurtin, M.E. and R.C. MacCamy: On the diffusion of biological populations Math. Biosci. 33 (1977), 35-49.
7. Namba, T.: Density-dependent dispersal and spatial distribution of a population, J. Theor. Biol. 86 (1980), 351-363.
8. Okubo, A.: Diffusion and ecological problems: mathematical models. Biomathematics 10 (Springer, 1980).
9. Peletier, L.A. and A. Tesei: Global bifurcation and attractivity of stationary solutions of a degenerate diffusion equation. Report Department of Mathematics, II University of Rome (1985).

L.A. Peletier
Rijksuniversiteit te Leiden
Subfaculteit Der Wiskunde en Informatica
Postbus 9512
2300 RA Leiden
Nederland

A. Tesei
Dipartimento di Matematica
Seconda Universitá di Roma
Via O. Raimondo
00173 Roma
Italia

M PIERRE

Nonlinear fast diffusion with measures as data

The results described here follow from technics developed in collaboration with M. Herrero and P. Baras.

Let us consider the following Cauchy problem

$$\begin{cases} (1)_a \quad u \in L^1_{\ell oc}((0,\infty) \times \mathbb{R}^n),\ u \geq 0 \\ (1)_b \quad \dfrac{\partial u}{\partial t} - \Delta(u^m) = 0 \quad \text{in} \quad \mathcal{D}'((0,\infty) \times \mathbb{R}^n) \\ (1)_c \quad \text{"}u(0,\cdot) = \mu\text{"} \end{cases} \tag{1}$$

where m is a real number in (0,1) and μ a given nonnegative Radon measure on $\mathbb{R}^N$. We want to know whether (1) has a solution.

If one assumes that μ belongs to $L^1(\mathbb{R}^N)$, it is now rather well-known that the answer is positive. Actually the operator $u \mapsto -\Delta(u|u|)^{m-1})$ generates a nonlinear semigroup in $L^1(\mathbb{R}^N)$ which preserves positivity (see [5], [6]).

If μ is in $L^1_{\ell oc}(\mathbb{R}^N)$, it has also been proved by Herrero-Pierre [9] that (1) has a unique solution for all time and this, no matter how fast μ(x) grows as |x| is large. In that case, u is continuous from [0,∞) into $L^1_{\ell oc}(\mathbb{R}^N)$.

However, it has been noticed by Brézis-Friedman [8] that, if μ is the Dirac mass at the origin, then (1) has a solution if and only if

$$m > (N-2)^+/N. \tag{2}$$

In that case, the initial datum "u(0,·) = μ" is understood as follows:

$$\begin{cases} \text{For any continuous function } \varphi \text{ with compact support} \\ \text{in } \mathbb{R}^N \text{ (essential)} - \lim\limits_{t \downarrow 0} \displaystyle\int_{\mathbb{R}^N} \varphi u(t) = \int_{\mathbb{R}^N} \varphi \, d\mu. \end{cases} \tag{3}$$

Now a natural question arises: assuming m is less than or equal to the critical value $(N-2)^+/N$, what are the measures μ that are "good enough" to provide a solution for (1)? This is the main question we solve below together with its "converse", namely: assume that u satisfies $(1)_a$ and $(1)_b$; then does u have a trace at t = 0?

Let us start with a main a priori estimate from which most of the results will follow.

Let u satisfy $(1)_a$, $(1)_b$ and let φ be a "test-function" in $C_o^\infty(\mathbb{R}^N)$ and nonnegative. According to $(1)_b$, we have

$$\frac{d}{dt}\int_{\mathbb{R}^N} \varphi u(t) = \int u^m \Delta\varphi \quad \text{in} \quad \mathcal{D}'(0,\infty).$$

Using Hölder's inequality, we deduce

$$\left|\frac{d}{dt}\int_{\mathbb{R}^N} \varphi u(t)\right| = \left|\int \varphi u^m \frac{\Delta\varphi}{\varphi}\right| \leq \left[\int \varphi u\right]^m F(\varphi) \tag{4}$$

where

$$F(\varphi) = \left[\int_{\mathbb{R}^N} \varphi\left[\frac{|\Delta\varphi|}{\varphi}\right]^{1/(1-m)}\right]^{1-m}. \tag{5}$$

Integrating the differential inequality (4) we are led to

$$\text{a.e. } t, s > 0 \quad \left[\int_{\mathbb{R}^N} \varphi u(t)\right]^{1-m} \leq \left[\int_{\mathbb{R}^N} \varphi u(s)\right]^{1-m} + (1-m)\,|t-s|\,F(\varphi). \tag{6}$$

In order to understand the meaning of (6), one needs to understand better the structure of the functional F.

<u>LEMMA 1.</u> *Let* $\varphi = \psi^{2/(1-m)}$ *with* $\psi \in C_o^\infty(\mathbb{R}^N)$, $0 \leq \psi \leq 1$. *Then, there exists* $C = C(m,N)$ *such that*

$$F(\varphi) \leq C\left[\int_{\mathbb{R}^N} |\Delta\psi|^{1/(1-m)}\right]^{1-m}.$$

<u>Proof.</u> We have

$$\Delta\varphi = \frac{2}{1-m}\left[\psi^{\frac{2}{1-m}-1}\Delta\psi + \frac{1+m}{1-m}\psi^{\frac{2m}{1-m}}|\nabla\psi|^2\right],$$

so that

$$\varphi\left[\frac{|\Delta\varphi|}{\varphi}\right]^{1/(1-m)} \le C(m)\left[\varphi^{1/(1-m)}|\Delta\psi|^{1/(1-m)} + |\nabla\psi|^{2/(1-m)}\right].$$

From Gagliardo-Nirenberg's inequality we know that

$$\int_{\mathbb{R}^N}|\nabla\psi|^{2p} \le C(p,N)\left|\int_{\mathbb{R}^N}|\Delta\psi|^p\right| \cdot \|\psi\|_\infty^p \quad \forall p \in [1,\infty[.$$

The lemma follows using $0 \le \psi \le 1$.

Thus from (6), we will essentially retain the following

LEMMA 2. *Let* u *satisfy* $(1)_a$, $(1)_b$. *Then, for all* ψ *in* $C_o^\infty(\mathbb{R}^N)$ *with* $0 \le \psi \le 1$, *we have with* $p = 1/(1-m)$

$$\text{a.e. } t,s > 0 \quad \left[\int_{\mathbb{R}^N}\psi^{2p}\,u(t)\right]^{1-m} \le \left[\int_{\mathbb{R}^N}\psi^{2p}\,u(s)\right]^{1-m} + C|t-s|\left[\int_{\mathbb{R}^N}|\Delta\psi|^p\right]^{1-m}. \tag{7}$$

REMARK. This estimate was established by Herrero-Pierre in [9].

We will deduce a good deal of information from (7). But first we need to introduce the notion of capacity associated with the Sobolev space $W^{2,p}(\mathbb{R}^N)$. Actually, we will only need to know the meaning of "zero capacity" and more precisely the next characterization (see [10], [3] for more details and references).

DEFINITION. Let K be a compact subset of $\mathbb{R}^N$. Then, if $c_{2,p}(\cdot)$ denotes the capacity associated with $W^{2,p}(\mathbb{R}^N)$, $c_{2,p}(K) = 0$ if and only if there exists a sequence ψ_n in $C_o^\infty(\mathbb{R}^N)$ such that

(i) $0 \le \psi_n < 1$

(ii) $\psi_n \ge 1$ on K

(iii) $\int_{\mathbb{R}^N} |\Delta\psi_n|^p \to 0$ as $n \to \infty$

(iv) $\psi_n \to 0$ a.e. and the support of ψ_n stays in a fix ball.

<u>REMARK.</u> Note that $(\int_{\mathbb{R}^N} |\Delta\psi|^p)^{1/p}$ is "locally" equivalent to the norm of ψ in $W^{2,p}(\mathbb{R}^N)$.

<u>THEOREM 1.</u> *Let* u *satisfy* $(1)_a$, $(1)_b$. *Then* u *has a "trace" at* t = 0: *more precisely, there exists a unique nonnegative Radon measure* μ *such that* $u(0,\cdot) = \mu$ *in the sense of* (3).

Moreover, μ does not charge the sets of $c_{2,p}$-capacity zero (with $p = (1-m)^{-1}$), that is

$$K \text{ compact}, \quad c_{2,p}(K) = 0 \Longrightarrow \mu(K) = 0. \tag{8}$$

<u>REMARK.</u> A similar "trace" result holds when $m > 1$. However, it is rather more delicate to obtain (see [2]) and is true only on the whole space $\mathbb{R}^N$ while it is a "local" result for $m \in (0,1)$.

With respect to the Cauchy problem (1), (8) appears as a necessary condition for the datum μ. We will see below that this condition is surprisingly also sufficient.

A point is of $c_{2,p}$-capacity zero if and only if $p \leq N/2$, that is $m \leq (N-2)^+/N$. As a consequence, a Dirac mass is not an admissible datum in that case : this was Brezis-Friedman's result mentioned at the beginning. On the other hand, if $m > (N-2)^+/N$, then (8) is empty since only the empty set is of capacity zero.

If μ is concentrated on a compact set of Hausdorff dimension d with $0 < d < N-2/(1-m)$, for the same reason it is not admissible either (see [10] for the relations between $c_{2,p}$ and Hausdorff measures). We can also deduce that, in the limit case $m = (N-2)^+/N$, measures without atoms are not necessarily admissible (see [4] for a similar argument).

<u>Proof of Theorem 1.</u> If u satisfies $(1)_a$, $(1)_b$, then applying Lemma 2 with $s = s_o$ fixed and $t \in (0,s_o) \setminus Z$ where Z is negligible, we obtain that for all set $K \subset \mathbb{R}^N$

$$\sup_{t\in(0,s_o)\setminus Z} \int_K u(t) < \infty.$$

This provides a compactness of u(t) in the space of measures. More precisely, there exists a nonnegative Radon measure on $\mathbb{R}^N$ and a sequence t_n converging to 0 such that $u(t_n)$ converges to μ in the sense of (3).

To establish that the whole "sequence" u(t) converges to μ along $(0,s)\setminus Z$, assume that $\hat{\mu}$ is another measure associated with another sequence s_p tending to 0. Applying Lemma 2 with $t = t_n$, $s = s_p$ and passing to the limit yield that, for any ψ in $C_o^\infty(\mathbb{R}^N)$ with $0 \leq \psi \leq 1$.

$$\int_{\mathbb{R}^N} \psi^{2p}\, d\mu \leq \int_{\mathbb{R}^N} \psi^{2p}\, d\hat{\mu}.$$

But this implies $\mu \leq \hat{\mu}$ and $\mu = \hat{\mu}$ by symmetry. This proves the first part of the theorem.

Now let K be a compact subset of $\mathbb{R}^N$ with $c_{2,p}(K) = 0$. Apply Lemma 2 with t = 0 (it is possible now) and with ψ_n as in the definition of "$c_{2,p}(K) = 0$". Using the properties of such a sequence φ_n, we easily get at the limit

$$\int_K d\mu \leq \liminf_{n\to\infty} \int_{\mathbb{R}^N} \psi_n^{2p}\, d\mu = 0.$$

<u>THEOREM 2.</u> *Let* μ *be a nonnegative Radon measure on* $\mathbb{R}^N$ *satisfying* (8). *Then there exists a solution* u *of the problem*

$$\begin{cases} u \in C((0,\infty);\ L^1_{\ell oc}(\mathbb{R}^N)),\ u \geq 0 \\ \dfrac{\partial u}{\partial t} = \Delta u^m \ \text{ in } \mathcal{D}'((0,\infty)\times\mathbb{R}^N) \\ u(0,\cdot) = \mu \textit{ in the sense of } (3). \end{cases} \tag{9}$$

<u>REMARK.</u> Notice that no growth condition is required on μ. This is mainly due to the "local" estimates of Lemma 2. Indeed, assume one can solve (9) for a sequence of initial data μ_n <u>increasing</u> to μ and assume that $\mu_n \mapsto u_n$ is monotone (this property will actually follow by construction). Applying (7)

to u_n with $s = 0$, $t > 0$ shows that for any set $K \subset \mathbb{R}^N$

$$\forall t \leq T \quad \int_K u_n(t) \leq C(T,K).$$

Therefore u_n increases to some $u \in L^1_{\ell oc}((0,\infty)\ \mathbb{R}^N)$. One can pass to the limit in the equation in the sense of distribution to obtain that

$$\frac{\partial u}{\partial t} = \Delta u^m.$$

By Lemma 2, we also have for any test-function ψ

$$\left|\left[\int_{\mathbb{R}^N} \psi^{2p}\ u_n(t)\right]^{1-m} - \left[\int_{\mathbb{R}^N} \psi^{2p}\ d\mu_n\right]^{1-m}\right| \leq C(m,\psi)t.$$

This is preserved at the limit. Hence μ is the initial trace of u. We will see below how the continuity from $(0,\infty)$ into $L^1_{\ell oc}(\mathbb{R}^N)$ is also preserved.

To prove Theorem 2 we need two independent results that we state as lemmas.

LEMMA 3. *(see Baras-Pierre* [3]*). Let μ be a nonnegative measure satisfying* (8). *Then, μ is the increasing limit of a sequence of nonnegative measures μ_n such that*

(i) μ_n *is compactly supported*
(ii) $\mu_n \in W^{-2,1/m}(\mathbb{R}^N)$ *dual space of* $W^{2,p}(\mathbb{R}^N)$.

REMARK. Measures which belong to $W^{-2,1/m}$ are known to be "diffuse enough" to fulfil (8). This lemma says that, up to an increasing process, measures satisfying (8) have that property (see [3] for a proof and more comments).

According to the remark following Theorem 2, we are now reduced to prove existence of a solution u to (9) when μ is a nonnegative and compactly supported measure of $W^{-2,1/m}(\mathbb{R}^N)$, together with the monotonicity of $\mu \to u$. The method will consist in approximating μ by $\mu_n = \mu * p_n$, where ρ_n is sequence of mollifiers, and showing that the corresponding solution u_n forms a relatively compact sequence in $L^1_{\ell oc}((0,\infty) \times \mathbb{R}^N)$ which converges to a solution of (9). The monotonicity property, well-known for classical solutions, will be preserved at the limit.

We need the next fundamental a priori estimate.

LEMMA 4. *(see Aronson-Bénilan [1] and Bénilan-Crandall [7]). Let* u *be the classical solution of* (9) *with* $u(0,\cdot) = u_0$ *in* $L^1(\mathbb{R}^N) \cap L^\infty(\mathbb{R}^N)$. *Then*

$$\frac{\partial u}{\partial t} \leq \frac{1}{1-m} \; \frac{u}{t} \quad \text{on} \quad (0,\infty) \times \mathbb{R}^N. \tag{10}$$

REMARK. This estimate essentially comes from the homogeneity of the non-linear operator $u \mapsto -\Delta(u^m)$.

We are now ready for the proof of Theorem 2.

Proof of Theorem 2. As indicated above, we approximate μ by $\mu_n = \mu * \rho_n$. Let u_n be the corresponding solution. By Lemma 2, u_n is bounded in $L^\infty((0,T) \times L^1_{\ell oc}(\mathbb{R}^N))$ for all T. However, this is not sufficient to pass to the limit in the equation. To get compactness properties we will use (10); note that it can be rewritten as

$$t \mapsto t^{-p}\, u(t) \quad \text{is nonincreasing.} \tag{11}$$

Hence for $\alpha > 0$ and K bounded in $\mathbb{R}^N$, we have

$$\int_{(\alpha,T)\times K} \left|\frac{\partial}{\partial t}\,(t^{-p}u_n(t))\right| = \int_{(\alpha,T)\times K} -\frac{\partial}{\partial t}\,(t^{-p}u_n(t)) \leq \alpha^{-p} \int_K u_n(\alpha).$$

It follows from this that $\frac{\partial u_n}{\partial t}$ and hence Δu_n^m are uniformly bounded in $L^1_{\ell oc}((0,\infty) \times \mathbb{R}^N)$, A similar estimate proves that $\frac{\partial}{\partial t}\,(u_n^m)$ has the same property. Consequently, u_n^m is relatively compact in $L^1_{\ell oc}((0,\infty) \times \mathbb{R}^N)$. Up to a subsequence, we can thus assume that

- u_n converges a.e. to u $L^1_{\ell oc}((0,\infty) \times \mathbb{R}^N)$
- u_n^m converges in $L^1_{\ell oc}((0,\infty) \times \mathbb{R}^N)$ to u^m
- $\Delta(u_n^m)$ converges in $\mathcal{D}'((0,\infty) \times \mathbb{R}^N)$ to Δu.

The only difficulty is to show that $\frac{\partial u_n}{\partial t}$ converges to $\frac{\partial u}{\partial t}$ in $\mathcal{D}'(\mathbb{R}^N)$; actually, this is false in general.

It is at this point that we will use the extra assumption $\mu \in W^{-2,1/m}(\mathbb{R}^N)$, which insures that μ_n converges to μ not only in the sense of measures but in $W^{-2,1/m}(\mathbb{R}^N)$.

To prove that $\frac{\partial u_n}{\partial t}$ tends to $\frac{\partial u}{\partial t}$ in $\mathcal{D}'$, it is sufficient to prove that u_n converges in $L^1_{\ell oc}((0,T) \times \mathbb{R}^N)$ to u. We already know that u_n converges a.e. to u. Thus by the dominated convergence theorem, it is sufficient to bound u_n from above by a sequence w_n which itself converges in $L^1_{\ell oc}$.

For this, we first restrict outselves to the case $N \geq 3$ to be able to deal with the potential of $u_n(t)$, that is

$$v_n(t) = \frac{C_N}{|x|^{N-2}} * u_n(t)$$

where C_N is a constant chosen so that $-\Delta v_n(t) = u_n(t)$.

<u>LEMMA 5.</u> *Assume* $N \geq 3$. *Then the classical solution* u *of* (9) *with* $\mu = u_o \in L^1(\mathbb{R}^N) \cap L^\infty(\mathbb{R}^N)$ *satisfies*

$$u^m(t,x) \leq \frac{C}{t} v_o(x) \tag{12}$$

where v_o *is the potential of* u_o.

We postpone the proof of this lemma and continue. Since μ_n converges in $W^{-2,1/m}(\mathbb{R}^N)$, its potential v_n converges in $L^{1/m}(\mathbb{R}^N)$. Thus (12) applied to u_n yields the pointwise estimate we were looking for. Therefore the limit u is solution of the equation.

We argue as before using Lemma 2 (see Remark after Theorem 2) to check that the initial trace of u is the expected μ.

For the continuity from $(0,\infty)$ into $L^1_{\ell oc}(\mathbb{R}^N)$, note that since $u \in L^\infty((0,T) ; L^1_{\ell oc}(\mathbb{R}^N))$ and $t^{-p} u(t)$ is monotone, u(t) has a right - and a left - limit in $L^1_{\ell oc}(\mathbb{R}^N)$ at each $t > 0$. We use Lemma 2 again to prove that they are equal. The continuity follows.

Finally, we need to treat the cases $N = 1,2$. Since $-\Delta$ is not invertible, we first replace it by $\varepsilon-\Delta$, whose inverse is defined on $L^1(\mathbb{R}^N)$. The same method as above provides a solution of $\frac{\partial u_\varepsilon}{\partial t} = (\varepsilon-\Delta)\, u^m$ with $u_\varepsilon(0,\cdot) = \mu$. It can be easily verified that $\varepsilon \mapsto u_\varepsilon$ is nonincreasing. We then let ε decrease to 0; passing to the limit is easy because of the monotonicity properties.

Proof of Lemma 5. According to [1], we can assume that u is C^∞ for $t > 0$. We set

$$v(t) = \frac{C_N}{|x|^{N-2}} * u(t), \tag{13}$$

where C_N is the constant chosen so that $-\Delta v(t) = u(t)$. According to (9) and (13) we have

$$\frac{\partial v}{\partial t} = -u^m.$$

In particular, $t \to v(t)$ is decreasing and we have

$$\forall 0 < s < t \qquad v(s) - v(t) = \int_s^t u^m(\sigma)d\sigma.$$

However, by (11) we deduce

$$v(s) - v(t) \geq t^{-pm} u^m(t) \int_s^t \sigma^{pm}\, d\sigma$$

and letting $s \to 0$,

$$v(0) \geq v(t) + t\, u^m(t)/(pm+1).$$

REMARK. According to Theorems 1 and 2, the measures μ for which (1) is solvable are exactly those for which the "stationary" problem

$$u - \lambda\Delta u^m = \mu \tag{14}$$

is solvable (see [3]). A different phenomenon occurs when one goes from the stationary problem

$$u + \lambda(-\Delta u + u^\gamma) = \mu \tag{15}$$

to its "evolution version", namely

$$\frac{\partial u}{\partial t} - \Delta u + u^{\gamma} = 0, \quad u(0,\cdot) = \mu. \tag{16}$$

For (15), (8) is the exact necessary and sufficient condition for solvability when $\gamma = 1/m$ (actually, it is essentially the same situation as in (14)). However, a different capacity appears in (16), namely the one associated with $W^{2m,p}(\mathbb{R}^N)$ (see [4]).

REFERENCES

1. Aronson, D.G. and P. Bénilan: Régularité des solutions de l'équation des milieux poreux dans $\mathbb{R}^N$, C.R. Acad. Sci. Paris 288 (1979), 103-105.
2. Aronson, D.G. and L. Caffarelli: The initial trace of a solution of the porous medium equation, Trans. Amer. Math. Soc. 280 (1983), 351-366.
3. Baras, P. and M. Pierre: Singularités éliminables pour des équations semi-linéaires, Ann. Inst. Fourier (Grenoble) 34 (1984), 185-206.
4. Baras, P. and M. Pierre: Problèmes paraboliques semi-linéaires avec données mesures, Applicable Anal. 18 (1984), 111-149.
5. Bénilan, P., H. Brézis and M.G. Crandall: A semilinear elliptic equation in $L^1(\mathbb{R}^N)$, Ann. Scuola Norm. Sup. Pisa 2 (1975), 523-555.
6. Bénilan, P. and M.G. Crandall: The continuous dependence on φ of solutions of $u_t - \Delta\varphi(u) = 0$, Indiana Univ. Math. J. 30 (1981), 162-177.
7. Bénilan, P. and M.G. Crandall: Regularizing effects of homogeneous evolution equations. In "Contributions to Analysis and Geometry", D.N. Clark Ed., pp. 23-29 (J. Hopkins Univ. Press, 1981).
8. Brézis, H. and A. Friedman: Nonlinear parabolic equations involving measures as initial conditions, J. Math. Pures Appl. 62 (1983), 73-97.
9. Herrero, M. and M. Pierre: The Cauchy problem for $u_t = \Delta u^m$ when $0 < m < 1$, Trans. Amer. Math. Soc.
10. Meyers, N.G.: A theory of capacities for potentials of functions in Lebesgue classes, Math. Scand. 26 (1970), 255-292.

M. Pierre
Département de Mathématiques
Université de Nancy I
B.P. 239
54506 Vandoeuvre les Nancy Cedex
France

J-P PUEL

A compactness theorem in quasilinear parabolic problems and application to an existence result

We present here a compactness theorem for quasilinear parabolic problems which is the essential part of the paper by Boccardo-Murat-Puel [3] concerning existence results for quasilinear parabolic equations.

In Section 1 we state the main compactness results; in Section 2 we give an application to the proof of an existence result for a quasilinear parabolic equation; in Section 3 we give the proof of the compactness theorem.

1. Statement of the results

In order to explain the results clearly, we will present two versions of the compactness theorem, a simple one and a more general one.

We will denote by Ω a bounded open set in R^n and by Q the cylinder $\Omega \times]0,T[$, where T is a positive number.

1.1. Simple statement

THEOREM 1. *Suppose we have a family* $(u_\varepsilon)_{\varepsilon>0}$ *satisfying the following properties:*

(i) $\forall\varepsilon > 0,\ u_\varepsilon \in L^2(0,T;H^1_o(\Omega)) \cap L^\infty(Q)$;

(ii)
$$\begin{cases} \dfrac{\partial u_\varepsilon}{\varepsilon t} - \Delta u_\varepsilon = g_\varepsilon & \text{in } Q, \\ u_\varepsilon(0) = u_0 & \text{in } \Omega, \end{cases} \tag{1.1}$$

where $u_0 \in L^\infty(\Omega)$, $g_\varepsilon \in L^1(Q)$, *and there exist a constant* C *and a function* $k \in L^1(Q)$ *such that*

$$\forall\varepsilon > 0,\ |g_\varepsilon(x,t)| \leq k(x,t) + C|\nabla u_\varepsilon(x,t)|^2 \quad \text{a.e. in } Q; \tag{1.2}$$

(iii) $\exists M > 0$ *such that*

$$\forall\varepsilon > 0,\ \|u_\varepsilon\|_{L^\infty(Q)} < M. \tag{1.3}$$

Then $(u_\varepsilon)_{\varepsilon>0}$ *is relatively compact in* $L^2(0,T;H_o^1(\Omega))$ *for the strong topology.*

REMARKS.

1) The boundedness of the family $(u_\varepsilon)_{\varepsilon>0}$ (and therefore the relative compactness for the weak topology in $L^2(0,T;H_o^1(\Omega))$) will be proved in a simple way but is not trivial.
2) Our proof does not require any regularity argument, so no regularity assumption has been made on Ω and we could consider various types of boundary conditions.

1.2. General statement.

Before giving the result, we need some notations and hypothesis.

Let $a_{ij}(x,t,s)$ be Caratheodory functions defined on $\Omega \times]0,T[\times R$ with values in R, such that:

$$a_{ij}(x,t,s) \text{ is uniformly bounded for bounded } s. \tag{1.4}$$

$$\exists\, \alpha > 0, \quad \forall \xi \in R^n, \ \forall s \in R, \quad \text{a.e. in } Q, \tag{1.5}$$

$$\sum_{i,j=1}^{n} a_{ij}(x,t,s)\, \xi_i\, \xi_j \geq \alpha|\xi|^2.$$

If u is a function defined on Q, we will write $A_{ij}(u)$ for the function defined by

$$A_{ij}(u)(x,t) = a_{ij}(x,t,u(x,t)) \quad \text{a.e. in } Q.$$

The matrix $(A_{ij}(u))$ will be denoted by A(u), and we also define the operator A(u) by

$$A(u) = \sum_{1,j=1}^{n} \frac{\partial}{\partial x_i} (A_{ij}(u) \frac{\partial u}{\partial x_j}) = -\nabla \cdot [A(u)\, \nabla u]. \tag{1.6}$$

THEOREM 2. *Let* $(u_\varepsilon)_{\varepsilon>0}$ *be a family such that*

(i) $\forall \varepsilon > 0, \quad u_\varepsilon \in L^2(0,T;H_o^1(\Omega)) \cap L^\infty(Q)$;

$$\text{(ii)} \begin{cases} \dfrac{\partial u_\varepsilon}{\partial t} + A(u_\varepsilon) = g_\varepsilon & \text{in } Q, \\ u_\varepsilon(0) = u_o & \text{in } \Omega, \end{cases} \tag{1.7}$$

where $u_o \in L^\infty(\Omega)$, $g_\varepsilon \in L^1(Q)$ *and*

$$\exists C > 0,\ \exists k \in L^1(Q) \text{ such that, } \forall \varepsilon > 0, \tag{1.8}$$

$$|g_\varepsilon(x,t)| \leq k(x,t) + C|\nabla u_\varepsilon(x,t)|^2, \quad \text{a.e.} \quad \text{in } Q;$$

$$\text{(iii)}\ \exists M > 0,\ \forall \varepsilon > 0 \quad \|u_\varepsilon\|_{L^\infty(Q)} \leq M\ .$$

Then $(u_\varepsilon)_{\varepsilon>0}$ *is relatively compact in* $L^2(0,T;H_o^1(\Omega))$ *for the strong topology.*

REMARKS.

1) Again here we need no regularity assumption on Ω nor on the coefficients a_{ij}, and we could have taken more general boundary conditions.
2) The result has been recently extended by Mokrane [6] to the case of an operator of Leray-Lions type in $L^p(0,T;W_o^{1,p}(\Omega))$ and with g_ε satisfying

$$\exists C > 0,\ \exists k \in L^1(Q),\ \forall \varepsilon > 0,\ |g_\varepsilon(x,t)| \leq k(x,t) + C|\nabla u_\varepsilon(x,t)|^p \text{ a.e. in } Q. \tag{1.8'}$$

   Then, of course the family $(u_\varepsilon)_{\varepsilon>0}$ is relatively compact in $L^p(0,T;W_o^{1,p}(\Omega))$.
3) An analogous compactness result for quasilinear elliptic equations has been obtained in Boccardo-Murat-Puel [2].

2. Application to quasilinear equations

Let $f(x,t,s,\xi)$ be a Caratheodory function defined on $\Omega \times]0,T[\times R \times R^n$ with values in R such that

$$|f(x,t,s,\xi)| \leq C(|s|)(1 + |\xi|^2) \quad \text{a.e.} \quad \text{in } Q, \tag{2.1}$$

where $r \to C(r)$ is an increasing function defined on R^+.

If u is a function defined on Q, we will denote by $F(u, \nabla u)$ the function defined by $F(u, \nabla u)(x,t) = f(x,t,u(x,t), \nabla u(x,t))$.

We are interested in quasilinear parabolic equations of the following type:

$$\begin{cases} \frac{\partial u}{\partial t} + A(u) + F(u, \nabla u) = 0 & \text{in } Q \\ u = 0 \quad \text{on} \quad \partial\Omega \times]0,T[\\ u(0) = u_o & \text{in } \Omega. \end{cases} \tag{2.2}$$

In order to apply Theorem 2 to get an existence result for (2.2), the main problem is to be able to define an "approximate" equation of the form

$$\begin{cases} \frac{\partial u_\varepsilon}{\partial t} + A(u_\varepsilon) + F_\varepsilon(u_\varepsilon, \nabla u_\varepsilon) = 0 & \text{in } Q \\ u_\varepsilon = 0 \quad \text{on} \quad \partial\Omega \times]0,t[\\ u_\varepsilon(0) = u_o & \text{in } \Omega, \end{cases} \tag{2.3}$$

such that, for every $\varepsilon \in]0, \varepsilon_o[$, (2.3) has a solution $u_\varepsilon \in L^2(0,T;H_o^1(\Omega)) \cap L^\infty(Q)$ satisfying:

$$\|u_\varepsilon\|_{L^\infty(Q)} \leq M \quad (M \text{ independent of } \varepsilon); \tag{2.4}$$

$$\exists k \in L^1(Q),\ \exists C > 0,\ |F_\varepsilon(u_\varepsilon, \nabla u_\varepsilon)(x,t)| \leq k + C|\nabla u_\varepsilon(x,t)|^2 \text{ a.e. in } Q. \tag{2.5}$$

Then, applying Theorem 2, it is easy to pass to the limit in (2.3).

Different sets of assumptions can be given in order to obtain a good approximate equation (2.3). One possibility (among others) is to assume the existence of a subsolution and a supersolution.

DEFINITION. • A function φ is a subsolution for (2.2) if

(i) $\varphi \in L^\infty(0,T;W^{1,\infty}(\Omega))$; $\frac{\partial \varphi}{\partial t} \in L^1(Q)$;

$$\text{(ii)} \begin{cases} \frac{\partial \varphi}{\partial t} + A(\varphi) + F(\varphi, \nabla\varphi) \leq 0 & \text{in } Q, \\ \varphi \leq 0 \quad \text{on} \quad \partial\Omega \times]0,T[, \\ \varphi(0) \leq u_o & \text{in } \Omega. \end{cases}$$

• A function ψ is a supersolution for (2.2) if

(i) $\psi \in L^\infty(0,T;W^{1,\infty}(\Omega))$; $\frac{\partial\psi}{\partial t} \in L^1(Q)$;

$$(ii) \begin{cases} \frac{\partial\psi}{\partial t} + A(\psi) + F(\psi, \nabla\psi) \geq 0 & \text{in } Q, \\ \psi \geq 0 \quad \text{on} & \partial\Omega \times]0,T[, \\ \psi(0) \geq u_o & \text{in } \Omega. \end{cases}$$

We then obtain the following existence result, using truncation arguments (see Boccardo-Murat-Puel [3]).

THEOREM 3. *If there exist a subsolution φ and a supersolution ψ for* (2.2) *such that $\varphi \leq \psi$ a.e. in Q, then there exists a solution of* (2.2) *satisfying*

$$\begin{cases} u \in L^2(0,T;H_o^1(\Omega)) \cap L^\infty(Q), \\ \frac{\partial u}{\partial t} \in L^1(Q) + L^2(0,T;H^{-1}(\Omega)), \\ \varphi \leq u \leq \psi \quad \text{a.e.} \quad \text{in } Q. \end{cases}$$

COMMENTS. Existence results for problems analogous to (2.2) have been obtained in several papers, but with additional assumptions concerning the regularity (of the functions, the set Ω and the boundary conditions) or the growth of f with respect to ξ (assuming the growth to be strictly less than quadratic). Among these papers we can mention Deuel-Hess [4], Amann [1], Puel [7], Frehse [5]. (See also the references in these papers).

Our result seems optimal concerning the regularity and growth assumptions, except for the regularity of the sub and supersolutions.

The proof that we give is completely different from the proofs given in the above mentioned papers, and it does not use any sophisticated argument.

3. Proof of the compactness theorem

We briefly give the proof of Theorem 2; it requires some tricks that are not necessary for Theorem 1.

3.1. Estimate in $L^2(0,T;H_o^1(\Omega))$.

We know that u_ε satisfies

$$\begin{cases} \dfrac{\partial u_\varepsilon}{\partial t} + A(u_\varepsilon) = g_\varepsilon, \\ u_\varepsilon(0) = u_o, \end{cases} \tag{3.1}$$

with (3.2) $|g_\varepsilon(x,t)| \le k(x,t) + C|\nabla u_\varepsilon(x,t)|^2$ a.e. in Q.

Let us consider the function

$$v_\varepsilon = e^{\lambda u_\varepsilon^2} \cdot u_\varepsilon = \varphi_\lambda(u_\varepsilon),$$

where λ is a positive number which will be chosen large enough, depending only on C and α.

Then

$$\frac{\partial v_\varepsilon}{\partial x_i} = e^{\lambda u_\varepsilon^2} \cdot \frac{\partial u_\varepsilon}{\partial x_i} + 2\lambda\, u_\varepsilon^2 \cdot e^{\lambda u_\varepsilon^2} \cdot \frac{\partial u_\varepsilon}{\partial x_i} ,$$

and v_ε belongs to $L^2(0,T;H_o^1(\Omega)) \cap L^\infty(Q)$ and is bounded in $L^\infty(Q)$. We multiply (3.1) by v_ε to get

$$\int_Q \frac{\partial u_\varepsilon}{\partial t}\varphi_\lambda(u_\varepsilon) + \sum_{i,j=1}^{n} \int_Q A_{ij}(u_\varepsilon)\frac{\partial u_\varepsilon}{\partial x_j} \cdot \frac{\partial u_\varepsilon}{\partial x_i} e^{\lambda u_\varepsilon^2}$$

$$+ 2\lambda \sum_{i,j=1}^{n} \int_Q A_{ij}(u_\varepsilon)\frac{\partial u_\varepsilon}{\partial x_j}\frac{\partial u_\varepsilon}{\partial x_i} \cdot e^{\lambda u_\varepsilon^2} \cdot u_\varepsilon^2 \le$$

$$\le \int_Q k|\varphi_\lambda(u_\varepsilon)| + C\int_Q |\nabla u_\varepsilon|^2 e^{\lambda u_\varepsilon^2} |u_\varepsilon| .$$

If $\Phi_\lambda(s) = \int_0^s \varphi_\lambda(\tau)d\tau$, we obtain

$$\Phi_\lambda(u_\varepsilon(T)) - \Phi_\lambda(u_o) + \alpha\int_Q e^{\lambda u_\varepsilon^2} |\nabla u_\varepsilon|^2 + 2\lambda\, \alpha\int_Q e^{\lambda u_\varepsilon^2} \cdot u_\varepsilon^2 |\nabla u_\varepsilon|^2$$

$$\le K + C\int_Q [e^{1/2\lambda u_\varepsilon^2} |\nabla u_\varepsilon|] \cdot [e^{1/2\lambda u_\varepsilon^2} |u_\varepsilon|\, |\nabla u_\varepsilon|]$$

$$\le K + \frac{\alpha}{2}\int_Q e^{\lambda u_\varepsilon^2} |\nabla u_\varepsilon|^2 + \frac{C^2}{2\alpha}\int_Q e^{\lambda u_\varepsilon^2} u_\varepsilon^2 |\nabla u_\varepsilon|^2 .$$

As $\Phi_\lambda(s) \geq 0$, choosing λ large enough ($\lambda \geq \frac{C^2}{4\alpha^2}$) we show that u_ε is bounded in $L^2(0,t;H_o^1(\Omega))$.

3.2. <u>Strong convergence in $L^2(Q)$.</u>

As u_ε is bounded in $L^2(0,T;H_o^1(\Omega))$, we can extract a subsequence denoted by $(u_m)_{m\in N}$ such that

$$u_m \longrightarrow u \quad \text{in } L^2(0,T;H_o^1(\Omega)) \quad \text{weakly.}$$

We know that u_m is bounded in $L^\infty(Q)$ and by (3.1), (3.2) that $\frac{\partial u_m}{\partial t}$ is bounded in $L^1(Q) + L^2(0,T;H^{-1}(\Omega))$, hence in $L^1(0,T;H^{-s}(\Omega))$ for s large enough.

Then, using a compactness lemma of Aubin's type (see Temam [8], Theorem 13.2), we can show that (u_m) is relatively compact in $L^2(0,T;H^{-s}(\Omega))$. But (u_m) is bounded in $L^2(0,T;H_o^1(\Omega))$. Then, using for instance an interpolation argument, (u_m) is relatively compact in $L^2(0,T;L^2(\Omega))$. Therefore

$$u_m \to u \quad \text{in } L^2(Q) \text{ strongly,}$$

and after extracting eventually a new subsequence, we can suppose that

$$u_m \to u \quad \text{a.e. in } Q.$$

3.3. <u>Strong convergence in $L^2(0,T;H_o^1(\Omega))$.</u>

We are going to prove that actually

$$u_m \to u \quad \text{in } L^2(0,T;H_o^1(\Omega)) \text{ strongly.}$$

Again we consider the function

$$\varphi_\lambda(s) = e^{\lambda s^2} \cdot s$$

and we choose λ such that

$$\alpha\varphi'_\lambda - 8\, C|\varphi_\lambda| \geq \frac{\alpha}{2} \tag{3.2}$$

((3.2) is satisfied for λ large enough).

We rewrite the equation for u_m as follows:

$$\begin{cases} \frac{\partial}{\partial t}(u_m - u_p) - \nabla\cdot[A(u_m)\,\nabla(u_m - u_p)] + \frac{\partial u_p}{\partial t} - \nabla\cdot[A(u_m)\,\nabla u_p] = g_m, \\ u_m - u_p = 0 \quad \text{on } \partial\Omega \times]0,T[, \\ (u_m - u_p)(0) = 0. \end{cases} \tag{3.3}$$

We multiply (3.3) by $\varphi_\lambda(u_m - u_p)$ which belongs to $L^2(0,T;H_o^1(\Omega)) \cap L^\infty(Q)$, and we obtain, writing $\langle , \rangle$ for the duality between $[L^2(0,T;H^{-1}(\Omega)) + L^1(Q)]$ and $[L^2(0,T;H_o^1(\Omega)] \cap L^\infty(Q)]$,

$$\int_\Omega \Phi_\lambda(u_m(T) - u_p(T)) + \int_Q A(u_m)\cdot\nabla(u_m - u_p)\cdot\nabla(u_m - u_p)\cdot\varphi'_\lambda(u_m - u_p)$$

$$+ \left\langle \frac{\partial u_p}{\partial t} - \nabla\cdot[A(u_m)\,\nabla u_p],\ \varphi_\lambda(u_m - u_p)\right\rangle = \int_Q g_m\,\varphi_\lambda(u_m - u_p)$$

$$\leq \int_Q k|\varphi_\lambda(u_m - u_p)| + C\int_Q |\nabla u_m|^2\,|\varphi_\lambda(u_m - u_p)|$$

$$\leq \int_Q k|\varphi_\lambda(u_m - u_p)| + 2C\int_Q [|\nabla(u_m - u_p)|^2 + |\nabla u_p|^2]\,|\varphi_\lambda(u_m - u_p)|.$$

If I is the identity matrix, since Φ_λ is positive we obtain

$$\int_Q [A(u_m)\varphi'_\lambda(u_m - u_p) - 2C|\varphi_\lambda(u_m - u_p)|I]\nabla(u_m - u_p)\cdot\nabla(u_m - u_p) \tag{3.4}$$

$$+ \left\langle\frac{\partial u_p}{\partial t},\ \varphi_\lambda(u_m - u_p)\right\rangle + \int_Q A(u_m)\nabla u_p\cdot\nabla[\varphi_\lambda(u_m - u_p)]$$

$$\leq \int_Q k|\varphi_\lambda(u_m - u_p)| + 2C\int |\nabla u_p|^2\,|\varphi_\lambda(u_m - u_p)|.$$

For fixed p, we let m tend to $+\infty$. We know that

$$u_m \to u \begin{cases} \text{in } L^2(0,T;H_o^1(\Omega)) & \text{weakly;} \\ \text{in } L^\infty(Q) & \text{weak } *; \\ \text{a.e.} & \text{in } Q. \end{cases}$$

If $D_m^p = A(u_m)\varphi'_\lambda(u_m - u_p) - 2C|\varphi_\lambda(u_m - u_p)|\ I$, we see that $D_m^p \geq \frac{\alpha}{2} I$ because of (3.2) and (1.5) and $D_m^p \to D^p$ a.e. in Q (and in $[L^\infty(Q)]^{N^2}$ weak*.), where

$$D^p = A(u)\varphi_\lambda'(u - u_p) - 2C|\varphi_\lambda(u - u_p)| \; I.$$

We also know that

$$\varphi_\lambda(u_m - u_p) \to \varphi_\lambda(u - u_p) \quad \text{in } L^2(0,T;H_0^1(\Omega)) \quad \text{weakly}$$

and in $L^\infty(Q)$ weak*, because it stays bounded and converges almost everywhere.

Then, using the positive definiteness of D_m^p, we can take the limit inferior in (3.4) and obtain

$$\int_Q [A(u)\varphi_\lambda'(u - u_p) - 2C|\varphi_\lambda(u - u_p)|I]\nabla(u - u_p) \cdot \nabla(u - u_p)$$
$$+\left\langle \frac{\partial u_p}{\partial t}, \varphi_\lambda(u - u_p)\right\rangle + \int_Q A(u) \nabla u_p \cdot \nabla[\varphi_\lambda(u - u_p)]$$
$$\leq \int_Q k|\varphi_\lambda(u - u_p)| + 2C \int_Q |\nabla u_p|^2 \; |\varphi_\lambda(u - u_p)|.$$

Now we use the equation satisfied by u_p and the bound on g_p to get

$$\int_Q [A(u)\varphi_\lambda'(u - u_p) - 2C|\varphi_\lambda(u - u_p)|I] \nabla(u - u_p) \cdot \nabla(u - u_p)$$
$$+\int_Q [A(u) - A(u_p)] \nabla u_p \cdot \nabla[\varphi_\lambda(u - u_p)]$$
$$\leq 2 \int_Q k|\varphi_\lambda(u - u_p)| + 3C \int_Q |\nabla u_p|^2 \; |\varphi_\lambda(u - u_p)|$$
$$\leq 2 \int_Q k|\varphi_\lambda(u - u_p)| + 6C \int_Q [|\nabla(u - u_p)|^2 + |\nabla u|^2]|\varphi_\lambda(u - u_p)|.$$

Using the relation

$$\int_Q [A(u)-A(u_p)]\nabla u_p \cdot \nabla[\varphi_\lambda(u-u_p)] = \int_Q [A(u_p)-A(u)]\varphi_\lambda'(u-u_p)\nabla(u-u_p)\cdot\nabla(u-u_p)$$
$$+ \int_Q [A(u) - A(u_p)] \nabla u \cdot \nabla[\varphi_\lambda(u-u_p)],$$

we obtain (notice the change in the first term!):

$$\int_Q [A(u_p)\varphi_\lambda'(u - u_p) - 8C|\varphi_\lambda(u - u_p)|I] \nabla(u - u_p) \cdot \nabla(u - u_p) \tag{3.5}$$

$$+ \int_Q [A(u) - A(u_p)] \nabla u \cdot \nabla[\varphi_\lambda(u - u_p)]$$

$$\leq 2 \int_Q k|\varphi_\lambda(u - u_p)| + 6C \int_Q |\nabla u|^2 \, |\varphi_\lambda(u - u_p)|.$$

Now because of (3.2) and (1.5) we have:

$$A(u_p)\varphi_\lambda'(u - u_p) - 8C|\varphi_\lambda(u - u_p)| \geq \frac{\alpha}{2} I ,$$

then (3.5) gives

$$\frac{\alpha}{2} \int_Q |\nabla(u - u_p)|^2 + \int_Q [A(u) - A(u_p)]\nabla u \cdot \nabla[\varphi_\lambda(u - u_p)] \tag{3.6}$$

$$\leq 2\int_Q k|\varphi_\lambda(u - u_p)| + 6C \int_Q |\nabla u|^2 \, |\varphi_\lambda(u - u_p)|.$$

Now we let p tend to infinity; then

$$u_p \to u \quad \begin{cases} \text{in } L^2(0,T;H_0^1(\Omega)) & \text{weakly;} \\ \text{in } L^\infty(Q) & \text{weak*;} \\ \text{a.e.} & \text{in } Q \end{cases}$$

and

$$\varphi_\lambda(u - u_p) \to 0 \quad \begin{cases} \text{in } L^2(0,T;H_0^1(\Omega)) & \text{weakly;} \\ \text{in } L^\infty(Q) & \text{weak*.} \end{cases}$$

We can take the limit in (3.6) to get

$$\frac{\alpha}{2} \int_Q |\nabla(u - u_p)|^2 \to 0 \quad \text{as} \quad p \to +\infty,$$

which shows that u_p converges strongly to u in $L^2(0,T;H_0^1(\Omega))$.

REFERENCES

1. Amann, H.: Existence and multiplicity theorems for semilinear elliptic boundary value problems, Math. Z. 150 (1976), 281-295.

2. Boccardo, L., F. Murat and J.P. Puel: Résultats d'existence pour certains problèmes elliptiques quasilinéaires, Ann. Scuola Norm. Sup. Pisa Cl. Sci. 11 (1984), 213-235.

3. Boccardo, L., F. Murat and J.P. Puel: Existence de solutions pour certains problèmes paraboliques quasilinéaires (to appear).

4. Deuel, J. and P. Hess: Nonlinear parabolic boundary value problems with upper and lower solutions, Israel J. Math. 29 (1978), 92-104.

5. Frehse, J.: Conference at the "International Workshop on Nonlinear Variational Problems", Isola d'Elba, 1983 (and article to appear).

6. Mokrane, A.: Existence of bounded solutions of nonlinear parabolic equations (to appear).

7. Puel, J.P.: Existence, comportement à l'infini et stabilité dans certains problèmes quasilinéaires elliptiques et paraboliques d'ordre 2, Ann. Scuola Norm. Sup. Pisa Cl. Sci. 3 (1976), 89-119.

8. Temam, R.: Navier-Stokes equations (North-Holland, 1977).

J.-P. Puel
Laboratoire d'Analyse Numérique
Tour 55/65, 5ème étage
Université Paris VI
4, Place Jussieu
75252 Paris Cedex 05
France

M SCHATZMAN

From a plateau to a thin mountain in the middle of a plain: asymptotic behaviour of nonlinear equations with a convolution term

1. Description of the models

We describe in this work, which is partly joint with E. Bienenstock and B. Moore, a family of mathematical models of the development of selectivity in the visual cortex - a part of the central nervous system. It has been observed that at some stage of development, certain regions of neural tissue respond approximately to all stimuli, and at a later stage, they respond selectively. The models try to explain this phenomenon. For all other neurophysiological informations, the reader is referred to [2].

In our family of equations, the spatial variable can be as well the spatial coordinate of a cell or the coordinate of a stimulus; this spatial variable could be discrete, but we treat here the continuous case only. The unknown function, denoted u(x,t) must be understood as a response to a unit stimulus, or a density of fibres, and we are interested in its asymptotic behavior as time goes to infinity. The main ingredients of the model are:

- a linear convolution term $w * u$; the convolution kernel w depends only on space, it describes long distance interaction between cells, it is even and continuous;

- a nonlinear term in two parts : a local term, expressed by a variational inequality (or a multivalued operator) keeps u non-negative, and a global term keeps the mass of u, or some of its partial integrals bounded.

If a diffusive effect is assumed, one can add a parabolic term $-\varepsilon\Delta u$, where ε is small.

If we define

$$F(u) = (w*u - f(u))\, 1_{\Omega}, \tag{1}$$

where Ω is $\mathbb{T}^N = \mathbb{R}^N/\mathbb{Z}^N$ or an open bounded subset of $\mathbb{R}^N$, and 1_Ω the characteristic function of Ω, our models are of the form

$$\frac{\partial^+ u}{\partial t} = \begin{cases} F(u) \text{ if } u > 0 \text{ or if } F(u) \geq 0, \\ 0 \quad \text{otherwise.} \end{cases} \tag{2}$$

In the language of maximal monotone operators, we define

$$\beta(r) = \begin{cases} \emptyset & \text{if } r < 0, \\ \mathbb{R}^- & \text{if } r = 0, \\ 0 & \text{if } r > 0; \end{cases} \tag{3}$$

then (2) is equivalent to

$$\frac{\partial u}{\partial t} + \beta(u) \ni F(u), \quad \text{a.e. on } \Omega \times]0,\infty[, \tag{4}$$

and its parabolic perturbation is

$$\frac{\partial u}{\partial t} + \beta(u) - \varepsilon\Delta u \ni F(u), \quad \text{a.e. on } \Omega \times]0,\infty[. \tag{5}$$

An initial condition u_0 is given; it is greater than or equal to zero, and, in the numerical simulations, it will be a positive constant, plus a small random perturbation. For (4), no boundary conditions are needed, because there is no spatial differentiation; however, for (5), one has to take Dirichlet, or Neumann, or periodic boundary conditions, according to the biological assumptions.

We give three main examples of function f, which correspond to different biological phenomena:

EXAMPLE 1. Orientation selectivity for a single neuron:

$$f^1(u) = (\int u(x)\, dx)^2. \tag{6}$$

We shall prove that the most stable states are Dirac masses, under the condition that w has a strict positive maximum at 0; in such a state, the neuron has no response at all angle of stimuli, except at one, where it has an infinite response. So sharp a selectivity is not observed, and this is the reason for introducing (5).

EXAMPLE 2. The spatial variable is a pair (x,y) and

$$f^2(u) = (\int u(x,y)\, dx)^2. \tag{7}$$

The most stable stationary states are masses carried by curves $x \to y(x)$, if $w(.,y)$ has a strict positive maximum at 0.

This model describes the spatial distribution of orientation selectivity for a collection of neurons : x is the angle of the stimulus, and y is the location of the neuron.

<u>EXAMPLE 3.</u> The spatial variable is a pair (x,y), and Ω_x and Ω_y are identical. Then

$$f^3(u) = \left(\int u(x,y)\,dx\right)^2 + \left(\int u(x,y)\,dy\right)^2. \tag{8}$$

If $w(x,x) + w(y,y) > 2\,w(x,y)$ whenever $x \neq y$, the most stable states are constant mass densities along the diagonals $x \pm y$ = constant. This model describes retinotopy, i.e. the one-to-one cabling between the retina, and a suitable neural sheet in the visual cortex.

If, for a moment, we think the coordinates in the retina and the visual cortex x, respectively y, as discrete variables, $u(x,y)$ is a transfer matrix: the response at y to a unit stimulus at x. An arbitrary positive matrix can evolve into a multiple of the identity matrix (up to some elementary geometric transformations).

Sharp selectivity creates the same problems for Example 2 and 3 as for Example 1; diffusion is an interesting way to keep the same general features with bounded u of small support.

2. Existence theory

The existence theory for (4) is given in an abstract setting : let V be a Banach space of measurable functions, ordered by

$$u \leq v \Leftrightarrow u(x) \leq v(x) \quad \text{a.e.} \tag{9}$$

It is assumed that, for every pair u,v of elements of V, $\max(u,v)$ belongs to V; moreover, the norm $\|\ \|$ is compatible with the order:

$$0 \leq u \leq v \Rightarrow \|u\| \leq \|v\|. \tag{10}$$

For each of the examples we considered, we have a different space

$V : V^1 = L^1(\Omega)$, $V^2 = L^\infty(\Omega_y; L^1(\Omega_x))$, and for Example 3, $V^3 = L^\infty(\Omega_y;L^1(\Omega_x)) \times \times L^\infty(\Omega_x;L^1(\Omega_y))$.

Let L, respectively M, be a linear continuous operator from V into itself, such that

$$u \leq v \Rightarrow Mu \leq Mv, \text{ and } u \geq 0 \Rightarrow |Lu| \leq Mu.$$

In all three examples, L is convolution by w, and M is convolution by $|w|$.

Finally, f is a function from V to itself which is locally Lipschitz continuous, and transforms non-negative functions into non-negative functions.

Then, we can state an abstract existence result:

THEOREM 1. *Under the above assumptions on* L *and* h, *for any nonnegative* u_o *in* V, *the problem*

$$u_t + f(u) + \beta(u) \ni Lu \tag{11}$$

with β *as in* (3) *possesses a unique solution in* $C^0([0,\infty);V)$, *such that* $\frac{\partial u}{\partial t}$ *belongs to* $L^\infty_{loc}([0,\infty);V)$; *moreover, we have*

$$\frac{\partial^+ u}{\partial t} = \begin{cases} [f(u) - Lu](x,t) \text{ if } u(x,t) > 0 \text{ or if } [f(u) - Lu](x,t) \geq 0, \\ \\ 0 \quad \text{otherwise,} \end{cases} \tag{12}$$

almost everywhere on $\Omega \times (0,\infty)$.

Outline of the proof. Consider the multivalued ordinary differential equation in $\mathbb{R}$, with a locally integrable right hand-side:

$$u_t + \beta(u) \ni f, \ u(0) = u_o \geq 0. \tag{13}$$

This problem possesses a unique continuous solution (which can be given explicitly), denoted by

$$u(t) = S(u_o,f,t).$$

The mapping $f \to du/dt$ is a contraction, and $du/dt \leq f$; these properties can be proved from the explicit expression given in [2], or from the techniques of [3], which are much more general. Let B_ρ be the ball of radius ρ around u_o; define a mapping T by

$$(Tu)(t) = S(u_o(.),h(u)(.),t); \tag{15}$$

we look for a fixed point of T in $C^o([0,\tau],B_\rho)$, for sufficiently small ρ and τ. This gives local existence; for global existence, observe that

$$\partial u/\partial t \leq Lu \leq Mu,$$

and, by a Gronwall technique,

$$u(t) \leq e^{Mt} u_o. \tag{16}$$

The pointwise interpretation of (11) is by the techniques of [3], slightly refined and modified to apply here.

Estimate (16) allows us to prove the existence of solutions of (11) in other spaces than the above defined V^j; see [2] for details. The existence theory for (5) is analogous, and somewhat easier, because there is more compactness.

3. Asymptotic behaviour

In all three cases considered here, there is a Lyapunov functional; it is given by

$$\Phi^1(u) = -\frac{1}{2}\int(w*u)\ u\ dx + \frac{1}{3}\left(\int u\ dx\right)^3, \tag{17}$$

$$\Phi^2(u) = -\frac{1}{2}\int(w*u)\ u\ dx\ dy + \frac{1}{3}\int\left(\int u(x,y)\ dx\right)^3 dy, \tag{18}$$

$$\Phi^3(u) = -\frac{1}{3}\int(w*u)\ u\ dx\ dy + \frac{1}{2}\int\left(\int u(x,y)\ dx\right)^3 dy + \tag{19}$$

$$+ \frac{1}{3}\int\left(\int u(x,y)\ dy\right)^3 dx.$$

When there is a diffusive term $- \varepsilon\Delta u$, the Lyapunov functionals are modified as follows:

$$\Phi^1_\varepsilon(u) = \Phi^1(u) + \varepsilon\int|\nabla u|^2 dx \text{ in case } 1, \tag{20}$$

$$\Phi^j_\varepsilon(u) = \Phi^j(u) + \varepsilon\int|\nabla u|^2 dx\, dy, \text{ for } j = 2 \text{ or } 3. \tag{21}$$

If we assume that w is bounded, the quadratic term can be estimated by the square of the norm in W_j, where

$$W^1 = V^1,\ W^2 = L^3(\Omega_y;\ L^1(\Omega_x)),\ W^3 = W^2 \cap L^3(\Omega_x);\ L^1(\Omega_y))$$

are spaces adapted to the respective Lyapunov functionals. If u belongs to W^j, u(t) remains bounded in W^j for all times, but the bounded subsets of W^j are not weakly relatively compact; therefore, we introduce super-spaces of the W^j's:

$$\tilde{W} = M^1(\Omega),\ \tilde{W}^2 = L^3(\Omega_y;\ M^1(\Omega_x)),$$

$$\tilde{W}^3 = \tilde{W}^2 \cap L^3(\Omega_x;\ M^1(\Omega_y)), \text{ with } M^1 \text{ space of measures.}$$

The bounded subsets of these new spaces are weakly* relatively compact; therefore, in Examples 1,2 and 3, the ω-limit set in the weak* topology of $\tilde{W}^j$ is not empty. We have a general theorem on the elements of the weak* ω-limit set:

<u>THEOREM 2.</u> *Let* u *belong to the ω-limit set of a trajectory of* (4); *then*

$$w*u \leq f^j(u);\ u \geq 0;\quad \operatorname{supp}(u) \subset \{w*u = f^j(u)\}.$$

Moreover, if u *belongs to the ω-limit set of* (5) *in* $W^j \cap H^1$, *then*

$$\varepsilon\Delta u + w*u \leq f^j(u) \text{ a.e.};\quad u \geq 0;$$
$$(\varepsilon\Delta u + w*u - f^j(u))\, u = 0 \quad \text{a.e.}\ .$$

Finally, we have results on the global minimizers of Φ^j over $\tilde{W}^j_+$, the set of non-negative elements of $\tilde{W}^j$.

THEOREM 3.

(i) *Assume that*

$$w(0) > 0, \ w(0) > w(x), \text{ for } x \neq 0;$$

then, any global minimizer of Φ^1 *over* $\tilde{W}^1_+$ *is a single Dirac mass* $u = w(0)\,\delta(.-a)$, *with a aribtrary in* $\mathbb{T}^N$.

Any sequence of global minimizers of Φ^1_ε *over* $H^1 \cap \tilde{W}^1_+$ *is weakly* relatively compact as* ε *tends to zero, and a converging subsequence tends to a global minimizer of* Φ^1.

(ii) *Assume that* $\Omega_x = \Omega_y = \mathbb{T}^N$ *and*

$$w(x,x) > 0, \ w(x,x) + w(y,y) > 2\,w(x,y), \text{ if } x \neq y;$$

then a global minimizer of Φ^3 *over* $\tilde{W}^3_+$ *is of the form*

$$u(x,y) = m\,\delta(x+y+a) \text{ or } m\,\delta(x-y+a),$$

where $m = \int w(x,x)\,dx/2$, *and a is arbitrary in* $\mathbb{T}^N$.

Numerical simulations for Examples 1 and 2 were given in [1]; in that article, there are simulations for a model of retinotopy with a different nonlinearity, which is only partially tractable mathematically.

The reader is referred to [2] for the proofs of the results of this paper.

REFERENCES

1. Bienenstock, E.: Cooperation and competition in the central nervous system development : a unifying approach. In "Synergetics", H. Haken ed. (Springer, to appear).

2. Bienenstock, E., M. Schatzman and B. Moore: Nonlinear systems with a spatial convolution term in a model of development of neuronal selectivity (to appear).

3. Brézis, H.: Operateurs maximaux monotones et semigroupes de contraction dans les espaces de Hilbert (North-Holland, 1973).

M. Schatzman
Mathématiques
Université Claude-Bernard
69 622 Villeurbanne Cedex
France

E SINESTRARI

An integrodifferential equation arising from the theory of nonlinear heat flow with memory

1. Introduction

The heat conduction in a homogeneous and isotropic body $\Omega \subset \mathbb{R}^3$ can be described through the internal energy $e(x,t)$, the heat flux $q(x,t)$ and the heat supply $s(x,t)$, which are related through the balance law for the heat transfer:

$$e_t(x,t) + \operatorname{div} q(x,t) = s(x,t). \tag{1.1}$$

If the dependence of e and q on the temperature $u(x,t)$ and its gradient $\nabla u(x,t)$ is known, one can derive from (1.1) an equation for the unknown $u(x,t)$. This dependence is given by the so-called constitutive relations, which depend on the physical properties of the body. If

$$e(x,t) = c_1 \cdot u(x,t), \tag{1.2}$$

$$q(x,t) = -c_2 \cdot \nabla u(x,t) \tag{1.3}$$

(with c_1, c_2 positive constants), we get from (1.1) the classical Fourier's law:

$$u_t(x,t) = c \cdot \Delta u(x,t) + f(x,t), \tag{1.4}$$

where $c = c_2/c_1$ and $f(x,t) = s(x,t)/c_1$.

The theory which can be deduced from this law is unable to account for memory effects; these are important for some materials, particularly at low temperatures. Thus we are led to consider more general constitutive assumptions, which take into account the material's thermal history, namely:

$$e(x,t) = \beta(u(x,t)) + \int_{-\infty}^{t} h(t-s)\gamma(u(x,s))ds, \tag{1.5}$$

$$q(x,t) = -\rho(\nabla u(x,t)) - \int_{-\infty}^{t} K(t-s)\sigma(\nabla u(x,s))ds. \tag{1.6}$$

Substituting into (1.1) gives the following equation for $u(x,t)$:

$$\frac{\partial}{\partial t}\beta(u(x,t)) + \frac{\partial}{\partial t}\int_{-\infty}^{t} h(t-s)\gamma(u(x,s))ds = \tag{1.7}$$

$$= \mathrm{div}\rho(\nabla u(x,t)) + \mathrm{div}\int_{-\infty}^{t} K(t-s)\sigma(\nabla u(x,s))ds + s(x,t).$$

With reference to this equation one can consider the following problem: given $u = \tilde{u}$ in $\Omega \times]-\infty,0]$, find u satisfying (1.7) in $\Omega \times [0,+\infty[$ with the homogeneous Dirichlet boundary condition

$$u = 0 \quad \text{in} \quad \partial\Omega \times [0,+\infty[; \tag{1.8}$$

here $\partial\Omega$ is the boundary of Ω (which is supposed to be bounded). In this case we get from (1.7)

$$\begin{cases} \dfrac{\partial}{\partial t}\beta(u(x,t)) + \dfrac{\partial}{\partial t}\displaystyle\int_{0}^{t} h(t-s)\gamma(u(x,s))ds = \\ \mathrm{div}\rho(\nabla u(x,t)) + \mathrm{div}\displaystyle\int_{0}^{t} K(t-s)\sigma(\nabla u(x,s))ds + f(x,t) \text{ in } \Omega\times[0,+\infty[, \end{cases} \tag{1.9}$$

where

$$f(x,t) = s(x,t) - \frac{\partial}{\partial t}\int_{-\infty}^{0} h(t-s)\gamma(\tilde{u}(x,s))ds +$$

$$\mathrm{div}\int_{-\infty}^{0} K(t-s)\sigma(\nabla\tilde{u}(x,s))ds.$$

Problem (1.8)-(1.9) has been considered by several authors under different conditions on the functions $\beta,\gamma,\rho,\sigma,f,h$ and k. Since the literature is relatively rich, we will limit ourselves to mention those papers which refer to the particular model studied in the sequel.

In 1968 M. Gurtin and A. Pipkin [12] proposed a non linear model of heat conduction with memory, which does not depend on the present value of the temperature gradient. When this theory is linearized, we get the heat flux relation

$$q(x,t) = -\int_{-\infty}^{t} K(t - s)\nabla u(x,s)ds. \tag{1.10}$$

Note that, if $K \equiv 1$, (1.9) can reduce to the classical wave equation; on the other hand, if k is the Dirac function, the memory effect vanishes in q and we can obtain also (1.4). We consider an intermediate case assuming that $K(t)$ has a weak singularity at $t = 0$. Concerning the other constitutive relation (1.5), we allow γ to be non linear and suppose h smooth in $[0,+\infty[$; for simplicity we also assume $\beta(u) = u$. Under these hypotheses (1.5) becomes

$$e(t,x) = u(x,t) + \int_{-\infty}^{t} h(t - s)\gamma(u(x,s))ds. \tag{1.11}$$

From (1.9), (1.10) and (1.11) we get:

$$\begin{cases} u_t(x,t) = \operatorname{div}\int_0^t K(t - s)\nabla u(x,s)ds - \\ \int_0^t h'(t - s)\gamma(u(x,s))ds - h(0)\gamma(u(x,t)) + f(x,t) \text{ in } \Omega \times [0,+\infty[. \end{cases} \tag{1.12}$$

We also prescribe the boundary and initial values for u, namely:

$$u = 0 \qquad \text{in } \partial\Omega \times [0,+\infty[\ ; \tag{1.13}$$

$$u = u_0 \qquad \text{in } \Omega \times \{0\}. \tag{1.14}$$

This problem can be given an abstract formulation by introducing a Banach space E of functions defined on Ω, the linear operator $A = \Delta$ with domain

$$D_A = \{u \in E; \ \ \Delta u \in E, \ u = 0 \ \text{ on } \Gamma\}$$

and the functions $u : [0,+\infty[\ \to E$, $\varphi : E \to E$, $f : [0,+\infty[\ \to E$ defined as follows:

$$u(t)(x) = u(x,t), \ \varphi(u)(x) = \gamma(u(x)), \ f(t)(x) = f(x,t).$$

With this notation, (1.12)-(1.14) can be written as a nonlinear integrodifferential equation in the Banach space E, namely:

$$\begin{cases} u'(t) = A\int_0^t K(t-s)u(s)ds - \int_0^t h'(t-s)\varphi(u(s))ds - \\ \qquad - h(0)\varphi(u(t)) + f(t) \qquad (t \geq 0) \\ u(0) = u_o \end{cases} \tag{1.15}$$

Obviously, the equivalence of (1.12)-(1.14) and (1.15)-(1.16) depends on the definition of solution of both problems; this point will be discussed later. Problem (1.15) is studied in [3] when E is a Hilbert space, $K(t) = -h'(t)$, $h(0) = 0$ and $K \in W^{2,1}_{loc}(0,+\infty)$: in this case a generalized solution is found in $[0,+\infty[$ when $x_\gamma(x) \geq 0$ for $x \in \mathbb{R}$.

Several authors ([1],[2],[4],[10],[13],[14],[16]-[19],[21]) have investigated problem (1.15) under various assumptions and with different methods; however, all of them assume the continuity of $K(t)$ at $t = 0$. The possibility of singular kernels is considered in [5]-[9], [11], whose results will be discussed later. After considering the case $\varphi = 0$ (i.e. the linear theory) we will investigate the semilinear equation (1.15); finally, an application to the heat conduction problem (1.12)-(1.14) will be given.

2. The linear problem

To treat the abstract problem (1.15) we consider first the case $\varphi = f = 0$:

$$\begin{cases} u'(t) = A\int_0^t K(t-s)u(s)ds \qquad (t > 0) \\ u(0) = u_o \end{cases} \tag{2.1}$$

and try to solve it heuristically by taking the Laplace transform of both members of the equation:

$$z\hat{u}(z) - u_o = A\,\hat{K}(z)\hat{u}(z).$$

From this we get

$$\hat{u}(z) = (z - \hat{K}(z)A)^{-1}u_o;$$

taking the inverse Laplace transform gives

$$u(t) = \frac{1}{2\pi i}\int_{+\Gamma} \frac{e^{zt}}{\hat{K}(z)}\left(\frac{z}{\hat{K}(z)} - A\right)^{-1} u_o dz, \tag{2.2}$$

where $+\Gamma$ is a suitable oriented curve in the complex plane.

We give now the exact assumptions on A and K in order to make the above arguments rigorous.

Let E be a Banach space, $A : D_A \subset E \to E$ a linear operator. Suppose there exist $M > 0$ and $\vartheta \in]\frac{\pi}{2}, \pi[$ such that:

$$\begin{cases} \text{if } z \in S_\vartheta = \{z \in \mathbb{C} \,;\, z \neq 0, \ |\operatorname{Arg} z| \leq \vartheta\}, \text{ then} \\ z \in \rho(A), \text{ the resolvent of } A, \text{ and we have} \\ \|(z - A)^{-1}\|_{L(E)} \leq \frac{M}{|z|} . \end{cases} \tag{2.3}$$

Concerning K we suppose that:

$$\begin{cases} K \in L^1_{loc}[0,+\infty); \quad \text{the Laplace transform } \hat{K}(z) \text{ is defined} \\ \text{and absolutely convergent for } \mathrm{Re}\, z > 0; \text{ moreover,} \\ \hat{K} \text{ has a holomorphic extension in a sector} \\ \\ S_\varphi = \{z \in \mathbb{C} \,;\, z \neq 0, \ |\arg z| \leq \varphi\} \\ \\ (\varphi \in]\frac{\pi}{2}, \pi[) \text{ so that the following condition is satisfied:} \\ \\ z \in S_\varphi \Rightarrow \hat{K}(z) \neq 0 \ \text{ and } \ \frac{z}{\hat{K}(z)} \in S_\vartheta \,; \end{cases} \tag{2.4}$$

$$\begin{cases} \text{there exist } c_1 > 0 \text{ and } \sigma < 1 \text{ such that, for each} \\ z \in S_\vartheta, \ |K(z)| \leq c_1 |z|^\sigma . \end{cases} \tag{2.5}$$

These assumptions will be always made in the sequel; as we shall see, they are satisfied in the applications to our heat conduction problem when K is a singular kernel. Let us first discuss how to define a resolvent operator T(t) for (2.1). More precisely, consider the curve $\Gamma = \Gamma_- \cup \Gamma_o \cup \Gamma_+$ where $\Gamma_\pm = \{z \in \mathbb{C};\ z = re^{\pm i\varphi},\ r \geq 1\}$ and $\Gamma_o = \{z \in \mathbb{C};\ z = e^{i\psi},\ |\psi| \leq \varphi\}$; set for each $t > 0$

$$T(t) = \frac{1}{2\pi i}\int_{+\Gamma} \frac{e^{zt}}{\hat{K}(z)}\left(\frac{z}{\hat{K}(z)} - A\right)^{-1} dz, \tag{2.6}$$

where $+\Gamma$ is oriented upwards. The following properties can be proved:

$$T(t) \in L(E) \text{ and } T(\cdot) \in C^{\infty}(]0,+\infty[;\, L(E)); \tag{2.7}$$

there exist $M_n (n = 0,1,\ldots)$ such that (2.8)

$$\|t^n T^{(n)}(t)\| \leq M_n \quad (t > 0);$$

$$\text{if } u_o \in \overline{D_A}, \text{ then } T(\cdot)u_o \in C([0,+\infty[;E). \tag{2.9}$$

If we suppose that K satisfies the additional assumption:

$$\begin{cases} \text{there exist } c_2 > 0 \text{ and } \rho \in \mathbb{R} \text{ so that for each } z \in S \\ |\hat{K}(z)|^{-1} \leq c_2 |z|^{\rho}, \end{cases} \tag{2.10}$$

then

$$T(t)u_o \in D_A \quad \text{for each } t > 0 \quad \text{and} \quad u_o \in E. \tag{2.11}$$

The proof of the previous and subsequent results can be found in [5,8,9], where it is supposed $\overline{D_A} = E$: however, their generalization to the case $\overline{D_A} \subset E$ is not difficult.

THEOREM 2.1. *Setting for* $u_o \in \overline{D_A}$:

$$u(t) = T(t)u_o, \tag{2.12}$$

we have $u \in C([0,+\infty[;E) \cap C^{\infty}(]0,+\infty[;E)$ *and*

$$u'(t) = A\int_0^t K(t - s)u(s)ds \quad (t > 0). \tag{2.13}$$

In other words, setting $T(0) = I$ we have that $u(t) = T(t)u_o \quad (t \geq 0)$ is a solution of the homogeneous problem (2.1). The uniqueness of such a solution is a consequence of a representation formula given by the following result.

THEOREM 2.2. *Suppose that* $f \in C(0,\overline{t};E)$ *and* $u_0 \in E$. *If there exists a solution* $u \in C([0,\overline{t}];E) \cap C^1(]0,\overline{t}];E)$ *of the problem*

$$\begin{cases} u'(t) = A\int_0^t K(t-s)u(s)ds + f(t) & (0 < t \leq \overline{t}) \\ u(0) = u_0 , \end{cases} \tag{2.14}$$

then we have necessarily

$$u(t) = T(t)u_0 + \int_0^t T(t-s)f(s)ds \qquad (0 \leq t \leq \overline{t}). \tag{2.15}$$

In [8] a condition is given on f, which ensures that u defined by (2.15) is in fact a solution of (2.14).

THEOREM 2.3. *If* $f \in C^\alpha([0,\overline{t}];E)$ *and* $u_0 \in \overline{D_A}$, *then* u *given by (2.15) is the unique solution* $u \in C([0,\overline{t}];E) \cap C^1(]0,\overline{t}];E)$ *of (2.14).*

It can be shown that the conclusion is true also if $f \in L^1(0,\overline{t};E) \cap C^\alpha([\varepsilon,\overline{t}];E)$ for each $\varepsilon \in]0,\overline{t}[$. It is useful now to introduce a family of subspaces D_α of E depending on a parameter $\alpha > 0$ (and on A, K):

$$D_\alpha = \{x \in E;\ \sup_{z \in S_\varphi} |z|^\alpha \left\|A\left(\frac{z}{\hat{K}(z)} - A\right)^{-1}x\right\| = [x]_\alpha < \infty\}, \tag{2.16}$$

endowed with the norm $\|x\|_{D_\alpha} = \|x\| + [x]_\alpha$.

It is easy to see that $D_\alpha \subset \overline{D_A}$ and $D_A \subset D_\alpha$ for $\alpha \leq 1 - \sigma$ (where σ is given in (2.5)); in addition, if $0 < \alpha < \alpha + \varepsilon$ we have $D_{\alpha+\varepsilon} \hookrightarrow D_\alpha$. The main property of these spaces is given by the following theorem.

THEOREM 2.4. *If* $x \in D_{n+\alpha}$ *with* $n = 0,1,\ldots$ *and* $0 < \alpha < 1$, *then* $T(\cdot)x \in C^{n+\alpha}([0,+\infty[;E)$.

For the proof see Theorem 3.3 of [9]. On the strength of this result we can not only examine the regularity of the solution of the homogeneous problem (2.1), but also give a maximal regularity result for the inhomogeneous problem (2.14) (see Theorem 4.4 of [9]).

THEOREM 2.5. *If* $f \in C^{\alpha}([0,\bar{t}];E)$, $f(0) \in D_{\alpha}$ *and* $u_o \in D_{1+\alpha}$, *then the solution of (2.14) belongs to* $C^{1+\alpha}([0,\bar{t}];E)$ *and the equation holds also for* $t = 0$.

Another case in which there exists a solution of (2.14) is when f takes values in some subspace D_{α} (see theorem 4.5 in [9]).

THEOREM 2.6. *Let* $f \in C([0,\bar{t}];E)$ *be such that* $\sup_{t\in[0,\bar{t}]} \|f(t)\|_{D_{\alpha}} < +\infty$ *for some* $\alpha \in]0,1[$; *let* $u_o \in D_{1+\varepsilon}$ *for some* $\varepsilon > 0$. *Then (2.14) has a solution* $u \in C([0,\bar{t}];E) \cap C^1(]0,\bar{t}];E)$.

A similar result has been proved in [11] where D_{α} is replaced by $D_{(-A)^{\alpha}}$, the domain of α-th power of $-A$.

3. A semilinear integrodifferential equation

Now we go back to problem (1.15), which can be treated by means of the results of the preceding section. Here we give only an application of the temporal regularity results.

THEOREM 3.1. *Let (2.3)-(2.5) hold and* $\varphi \subset Lip_{loc}(E,E)$. *If for some* $\alpha \in]0,1[$ *we have* $f \in C^{\alpha}([0,\bar{t}],E)$, $h \in W^{1,\frac{1}{1-\alpha}}(0,\bar{t})$ *and* $u_o \in D_{\alpha}$ *then there exists* $\tau > 0$ *and a unique* $u \in C^{\alpha}([0,\tau];E) \cap C^1(]0,\tau];E)$ *such that*

$$\begin{cases} u'(t) = A\int_0^t K(t-s)u(s)ds - \int_0^t h'(t-s)\varphi(u(s))ds - \\ \qquad h(0)\varphi(u(t)) + f(t) \quad (0 < t \leq \tau) \\ u(0) = u_o \end{cases} \tag{3.1}$$

If in addition $u_o \in D_{1+\alpha}$ *and* $f(0) - h(0)\varphi(u_o) \in D_{\alpha}$, *then we have* $u \in C^{1+\alpha}([0,\tau];E)$ *and the equation of (3.1) holds also for* $t = 0$.

Proof. Setting

$$\gamma_u(t) = -\int_0^t h'(t-s)\varphi(u(s))ds - h(0)\varphi(u(t)) + f(t) \tag{3.2}$$

we will prove the existence of a continuous function from $[0,\tau]$ to E such that

$$u(t) = T(t)u_0 + \int_0^t T(t-s)\gamma_u(s)ds \qquad (0 \leq t \leq \tau) \tag{3.3}$$

provided τ is sufficiently small.

Let r_0 and c_0 be such that

$$\|\varphi(x) - \varphi(y)\| \leq c_0\|x - y\|.$$

For $x,y \in B(u_0,r_0) = \{\xi \in E;\ \|\xi - u_0\| \leq r_0\}$ and for each $\tau \in]0,\bar{t}]$ let us define the Banach space

$$X_\tau = \{u \in C([0,\tau];E);\ u(t) \in B(u_0,r_0),\ t \in [0,\tau]\}$$

endowed with the supremum norm. It can be proved that if τ is sufficiently small the operator $u \to S(u)$ (where $S(u)(t)$ denotes the right-hand side of (3.3) is a contraction of X_τ into itself: hence there exists a unique solution $u \in X_\tau$ of (3.3). By inspection of the right-hand side of (3.3) and by virtue of Theorem 2.4 and of Lemma 4.1 of [9], we deduce that $u \in C^\alpha([0,\tau];E)$, thus $\gamma_u \in C^\alpha([0,\tau];E)$. Now from (3.3) and Theorems 2.3, 2.5 we get the conclusion.

4. Application to the heat flow problem

Let us set now

$$\begin{cases} E = C(\overline{\Omega}) \text{ with the supremum norm} \\ Au = \Delta u \\ D_A = \{u \in C(\overline{\Omega});\ u|_\Gamma = 0,\ \ \Delta u \in C(\overline{\Omega})\}, \end{cases} \tag{4.1}$$

where Δ is understood in the distributional sense. It is well-known that (2.3) is satisfied for any $\vartheta \in]\frac{\pi}{2},\pi[$ with a suitable $M = M(\vartheta)$ (see [22]).

Now if we set

$$K(t) = \frac{1}{t^\beta} \qquad (t > 0) \tag{4.2}$$

with $\beta \in]0,1[$, we have for $\mathrm{Re}\lambda > 0$

$$\hat{K}(\lambda) = \Gamma(1 - \beta)\lambda^{\beta-1};$$

thus (2.4) is satisfied if $2 - \frac{2\vartheta}{\pi} < \beta$ by choosing $\varphi = \frac{\vartheta}{2-\beta}$. Finally (2.5) and (2.10) hold with $c_1 = c_2^{-1} = c\Gamma(1-\beta)$ and $\rho = -\sigma = 1-\beta$. Let us recall now the definition of the real interpolation spaces between D_A and E depending on the parameter $\alpha \in]0,1[$ (see e.g. [20]):

$$D_A(\alpha,\infty) = \{x \in E;\ \sup_{\xi>0} \|\xi^{1-\alpha} A e^{A\xi} x\| = \|x\|_\alpha < \infty\} \tag{4.3}$$

with norm $\|x\| + \|x\|_\alpha$. These spaces have been characterized in [15] as follows:

$$D_A(\alpha,\infty) \simeq C_0^{2\alpha}(\overline{\Omega}) = \{u \in C^{2\alpha}(\overline{\Omega});\ u|_\Gamma = 0\} \tag{4.4}$$

for $\alpha \neq \frac{1}{2}$; the case $\alpha = \frac{1}{2}$ requires the introduction of other functional spaces and will not be considered for brevity. By using (4.3) we get for $0 < \alpha < 2 - \beta$

$$D_\alpha \simeq D_A\left(\frac{\alpha}{2-\beta},\infty\right). \tag{4.5}$$

Therefore we have in this case

$$D_\alpha \simeq C_0^{\frac{2\alpha}{2-\beta}}(\overline{\Omega}) \quad \text{for} \quad 0 < \alpha < 2 - \beta. \tag{4.6}$$

We can apply now Theorem 3.1 to get a solution of problem (1.12)-(1.14) when $K(t) = \frac{1}{t^\beta}$.

<u>THEOREM 4.1.</u> *Let* $\alpha > 0$, $\beta < 1$, $\alpha \neq 1-\beta/2$, $h \in W^{1,\frac{1}{1-\alpha}}(0,\overline{t})$, $\gamma \in C^1(\mathbb{R})$, $f(x,\cdot) \in C^\alpha(0,\overline{t})$ *uniformly for* $x \in \overline{\Omega}$ *and* $u_0 \in C_0^{\frac{2\alpha}{2-\beta}}(\overline{\Omega})$. *Then there exists* $\tau \in]0,\overline{t}]$ *and a unique* $u \in C(\overline{\Omega} \times [0,\tau])$ *such that* ∇u *and* u_t *are continuous in* $\overline{\Omega} \times]0,\tau]$ *and*

$$\begin{cases} u_t(x,t) = \operatorname{div}\displaystyle\int_0^t \frac{\nabla u(x,s)}{(t-s)^\beta}\,ds \; - \\ - \displaystyle\int_0^t h'(t-s)\gamma(u(x,s))ds - h(0)\gamma(u(x,t)) + f(x,t) \quad \text{in } \overline{\Omega} \times]0,\tau] \\ u = 0 \quad \text{in} \quad \partial\Omega \times]0,\tau[\\ u = u_0 \quad \text{in} \quad \overline{\Omega} \times \{0\}. \end{cases} \tag{4.7}$$

Proof. With the above notations, we can use the first part of Theorem 3.1 to show that

$$\Delta\int_0^t \frac{u(x,s)}{(t-s)^\beta}\,ds = \operatorname{div} \int_0^t \frac{\nabla u(x,s)}{(t-s)^\beta}\,ds \tag{4.8}$$

and $u(x,t) = 0$ when $(x,t) \in \partial\Omega \times]0,\tau[$. This can be proved using the fact that, if $f \in C(0,\overline{t};E)$ and $u_0 \in E$, then

$$u(t) = T(t)u_0 + \int_0^t T(t-s)f(s)ds$$

belongs to $C(]0,\overline{t}];D_A(\alpha',\infty)) \cap L^1([0,\overline{t}];D_A(\alpha',\infty))$ with some $\alpha' \in]\frac{1}{2}, 1[$; observe moreover that $D_A(\alpha',\infty) \subset \overline{D_A}$. Since $D_A(\alpha',\infty) \to C^1(\overline{\Omega})$, also (4.8) holds.

REFERENCES

1. Aizicovici, S.: On a nonlinear integrodifferential equation, J. Math. Anal. Appl. 63 (1978), 385-395.
2. Aizicovici, S.: Existence theorems for a class of integrodifferential equations, An. Stiint. Univ. "Al. I. Cuza" Jasi 24 (1978), 113-124.
3. Aizicovici, S.: On a semilinear Volterra integrodifferential equation, Israel J. Math. 36 (1980), 273-284.
4. Belleni-Morante, A.: An integrodifferential equation arising from the theory of heat conduction in rigid materials with memory, Boll. Un. Mat. Ital. B. 15 (1978), 470-482.
5. Da Prato, G. and M. Iannelli: Linear integro-differential equations in Banach spaces, Rend. Sem. Mat. Univ. Padova 62 (1980), 207-219.
6. Da Prato, G. and M. Iannelli: Linear abstract integrodifferential equations of hyperbolic type in Hilbert spaces, Rend. Sem. Mat. Univ. Padova 62 (1980), 191-206.

7. Da Prato, G. and M. Iannelli: Distribution resolvents for Volterra equations in a Banach space, J. Integral Equations 6 (1984), 93-103.

8. Da Prato, G., M. Iannelli and E. Sinestrari: Temporal regularity for a class of integrodifferential equations in Banach spaces, Boll. Un. Mat. Ital. Anal. Funz. 2 (1983), 171-185.

9. Da Prato, G., M. Iannelli and E. Sinestrari: Regularity of solutions of a class of linear integrodifferential equations in Banach spaces, J. Integral Equations 8 (1985), 27-40.

10. Davis, P.L.: On the hyperbolicity of the equations of the linear theory of heat conduction for materials with memory, SIAM J. Appl. Math. 30 (1976), 75-80.

11. Grimmer, R. and A.J. Pritchard: Analytic resolvent operators for integral equations in Banach space, J. Differential Equations 50 (1983), 234-259.

12. Gurtin, M.E. and A.C. Pipkin: A general theory of heat conduction with finite wave speeds, Arch. Rational Mech. Anal. 31 (1968), 113-126.

13. Londen, S.O.: An existence result on a Volterra equation in a Banach space, Trans. Amer. Math. Soc. 235 (1978), 285-304.

14. Londen, S.O.: On an integrodifferential Volterra equation with a maximal monotone mapping, J. Differential Equations 27 (1978), 405-420.

15. Lunardi, A.: Interpolation spaces between domains of elliptic operators and spaces of continuous functions with applications to nonlinear parabolic equations, Math. Nachr. 121 (1985), 295-318.

16. MacCamy, R. and J. Wong: Stability theorems for some functional equations. Trans. Amer. Math. Soc. 164 (1972), 1-37.

17. MacCamy, R.: Stability theorems for a class of functional differential equations, SIAM J.Appl. Math. 30 (1976), 557-576.

18. MacCamy, R.: An integrodifferential equation with application in heat flow, Quart. Appl. Math. 35 (1977), 1-19.

19. Miller, R.K.: An integrodifferential equation for rigid heat conductors with memory, J. Math. Anal. Appl. 66 (1978), 313-332.

20. Sinestrari, E.: On the abstract Cauchy problem of parabolic type in spaces of continuous functions, J. Math. Anal. Appl. 107 (1985), 16-66.

21. Staffans, O.: On a nonlinear hyperbolic Volterra equation, SIAM J. Math. Anal. 11 (1980), 793-812.

22. Stewart, H.B.: Generation of analytic semigroups by strongly elliptic operators, Trans. Amer. Math. Soc. 199 (1974), 141-162.

E. Sinestrari
Dipartimento di Matematica,
Università di Roma "LA SAPIENZA"
Piazza A. Moro
00185 Roma
Italia

G M TROIANIELLO

Structure of the solution set for a class of nonlinear parabolic problems

1. In the case of ordinary differential equations on a compact interval I the set (supposed nonvoid) of solutions to an initial value problem, if not a singleton, is compact and connected in the space of continuous functions on I. This is referred to as the "Hukuhara-Kneser property". Extensions of it to the case of parabolic equations have been given in the 60's by Krasnosel'skii and Sobolevskii [3], Bebernes and Schmitt [2].

2. Here we illustrate the results of [7] about the set $\Gamma(u_o)$ of functions $u \equiv (u_1,\ldots,u_n)\colon Q \to \mathbb{R}^n$ which satisfy

$$\begin{cases} \varphi_k \leq u_k \leq \psi_k, \\ [E_k u_k - F_k(u,\nabla u_k)](u_k - \varphi_k) \leq 0 \\ [E_k u_k - F_k(u,\nabla u_k)](u_k - \psi_k) \leq 0 \\ \text{in } Q \text{ for } k = 1,\ldots,n, \end{cases} \tag{1}$$

$$u = 0 \text{ on } \Sigma,\ u(\cdot,0) = u_o \text{ in } \Omega, \tag{2}$$

where Q is the Cartesian product between a bounded domain $\Omega \subset \mathbb{R}^N$ and a bounded interval $]0,t[$, Σ the Cartesian product between the boundary $\partial\Omega$ (supposed smooth) of Ω and $]0,t[$. Each $E_k v$ (for $v\colon Q \to \mathbb{R}$) is given by

$$E_k v \equiv v_t - \sum_{i,j=1}^{N} a_k^{ij} v_{x_i x_j}$$

with $a_k^{ij} \in C^o(\overline{Q})$ for $i,j = 1,\ldots,N$,

$$\sum_{i,j=1}^{N} a_k^{ij}\xi_i\xi_j > 0 \text{ on } \overline{Q} \text{ for } \xi \in (\xi_1,\ldots,\xi_N) \in \mathbb{R}^N \quad \{0\};$$

each $F_k(w,z)$ (for $w\colon Q \to \mathbb{R}^n$ and $z\colon Q \to \mathbb{R}^N$) is the Nemytsky operator associated with a Caratheodory function $f_k: \Omega \times]0,T[\times \mathbb{R}^n \times \mathbb{R}^N \to \mathbb{R}$, that is,

$$[F_k(w,z)](x,t) \equiv f_k(x,t,w(x,t),z(x,t)).$$

Note that, if the a_k^{ij}'s are regular enough, say Lipschitzian with respect to $x \in \overline{\Omega}$, problem (1)-(2) amounts to a system of parabolic variational inequalities of the bilateral type.

Solutions u to (1)-(2) are sought for in the N-th Cartesian power of $W_p^{2,1}(Q)$, the Sobolev space of measurable functions $v: Q \to \mathbb{R}$ which belong to $L_p(Q)$ together with their derivatives $v_{x_i}, v_{x_ix_j}$ $(i,j = 1,\dots,N)$, v_t; p is assumed $> N+2$.

3. As in the theory of ordinary differential equations, uniqueness of solutions to (1)-(2) is ensured, if the φ_k's and ψ_k's are simply assumed continuous, by a condition of the Lipschitz type on the functions $f_k(x,t,w,z)$ with respect to $w \in \mathbb{R}^n$ and $z \in \mathbb{R}^N$. But our interest concerns the case when the solution set $\Gamma(u_o)$ is not empty and contains more than one element. We suppose that, whatever $L \in]0,\infty[$,

$$|f_k(x,t,w,z)| \leq \gamma_L(x,t)+K_L|z|^2$$
for a.a. $(x,t) \in Q$,
any $w \in \mathbb{R}^n$ with $|w| \leq L$, any $z \in \mathbb{R}^N$,

where $\gamma_L \in L_p(Q)$ and $0 \leq K_L < \infty$, and that

$$\varphi_k = \bigvee_{h=1}^{m} \varphi_{k,h}$$

with $\varphi_{k,h} \in W_p^{2,1}(Q)$, $\varphi_{k,h} \leq 0$ on Σ,

$$\psi_k = \bigwedge_{h=1}^{m} \psi_{k,h}$$

with $\psi_{k,h} \in W_p^{2,1}(Q)$, $\psi_{k,h} \geq 0$ on Σ,

$$\varphi_k \leq \psi_k \quad \text{on } \overline{Q}$$

$(k = 1,\dots,n)$. Denoting by $W_p^{2-2/p}(\Omega)$ the Sobolev space of fractionary order $2-2/p$ and exponent p on Ω, and by $\mathbb{K}_o$ the set of functions $v_o \equiv (v_{o1},\dots,v_{on}) \in [W_p^{2-2/p}(\Omega)]^n$ satisfying

$$v_{ok} = 0 \text{ on } \partial\Omega, \ \varphi_k(\cdot,0) \leq v_{ok} \leq \psi_k(\cdot,0) \text{ in } \Omega$$

for $k = 1,\dots,n$, we have

THEOREM 1. *Under the above assumptions about the functions* f_k, φ_k, ψ_k *the solution set* $\Gamma(u_o)$ *is not empty whatever* $u_o \in \mathbb{K}_o$.

(For $n = 1$, even if the leading coefficients of the parabolic operator are assumed to depend continuously on $u \in \mathbb{R}$ in addition to $(x,t) \in \overline{Q}$, we can improve Theorem 1 by showing that $\Gamma(u_o)$ admits a maximal and a minimal element: see [6]).

4. Simple examples show that, under the general assumptions of Theorem 1, uniqueness need not hold. To compensate for the lack of uniqueness we now give sufficient conditions in order that $\Gamma(u_o)$, even if not a singleton, is compact and "more than" connected in

$$\mathbb{X} \equiv C^o([0,t]; [W_p^{2-2/p}(\Omega)]^n).$$

The latter point needs some clarifying comment. In algebraic topology the connectedness of a topological space can be expressed through by requiring that its reduced cohomology group of order 0 is trivial. What we can prove about $\Gamma(u_o)$, besides compactness, is that its reduced cohomology groups of all orders ≥ 0 are trivial -i.e., $\Gamma(u_o)$ is acyclic. We however avoid relying too heavily on algebraic topology and formulate our result as a purely analytic property, which could then be shown to imply not only connectedness but even acyclicity.

It would be important to improve the acyclicity result by showing that $\Gamma(u_o)$ is an R_δ-set (that is, the intersection of a decreasing sequence of compact absolute retracts), thus extending to parabolic operators a theorem by Aronszajn [1] concerning ordinary differential equations. To the best of our knowledge, this question remains open.

We need to strengthen the assumptions of Section 3. Namely we suppose that $f_k \in C^o(\overline{\Omega} \times [0,T] \times \mathbb{R}^n \times \mathbb{R}^N)$, that, whatever $L \in]0,\infty[$,

$$|f_k(x,t,w,z)| \leq K_L(1 + |z|^2)$$

for any $(x,t) \in \overline{Q}$,

any $w \in \mathbb{R}^n$ with $|w| \leq L$, any $z \in \mathbb{R}^N$,

and that all function $\varphi_{k,h}$, $\psi_{k,h}$ belong to $W^{2,1}_\infty(Q)$.

THEOREM 2. *Let* $U \subset \mathbb{K}_o$ *be bounded in* $W_p^{2-2/p}(\Omega)$. *Under the above assumptions about the functions* f_k, φ_k, ψ_k *the solution set* $\Gamma(u_o)$, $u_o \in \mathbb{U}$, *satisfies*

$$\Gamma(u_o) = \bigcap_{\mu=1}^{\infty} \Gamma^\mu(u_o),$$

where $\{\Gamma^\mu(u_o)\}$ *is a sequence (dependent on* $\mathbb{U}$*) of subsets of* $\mathbb{X}$ *with the following properties:*

(i) *for any* $\mu \in \mathbb{N}$ *there exist a nonvoid, convex and compact subset* B^μ *of some topological vector space, and a continuous singlevalued mapping*

$$\gamma^\mu : \mathbb{U} \times B^\mu \to \mathbb{X}$$

such that

$$\Gamma^\mu(u_o) = \gamma^\mu(u_o, B^\mu) \text{ for } u_o \in \mathbb{U}$$

and

$[\gamma^\mu(u_o,\cdot)]^{-1}(u)$ *is convex for* $u \in \gamma^\mu(u_o)$;

(ii) *for any* $\mu \in \mathbb{N}$,

$$\gamma^\mu(u_o) \supseteq \gamma^{\mu+1}(u_o) \text{ for } u_o \in \mathbb{U}.$$

Note that the Hukuhara-Kneser property of $\Gamma(u_o)$ follows from the fact that $\Gamma^{\mu}(u_o)$ is the image of the compact and connected topological space B^{μ} under the mapping $\gamma^{\mu}(u_o,\cdot)$.

5. The scope of the analytic properties provided by Theorem 2 appears more clearly in the proof of our next result, which utilizes such properties to find a fixed point for a multivalued Poincaré map. This could not be achieved if the $\Gamma(u_o)$'s were only compact and connected. (For a different approach, based on algebraic topology, see [4]).

We pass from initial-boundary conditions to periodic boundary conditions, that is, from (2) to

$$u = 0 \quad \text{on} \quad \Sigma,\ u(\cdot,0) = u(\cdot,T) \quad \text{in} \quad \Omega. \tag{2)'$$

Solutions to (1)-(2)' are fixed point of the multivalued mapping $\Pi\circ\Gamma$, with $\Pi u \equiv u(\cdot,T)$ for $u \in \mathbb{X}$.

<u>THEOREM 3.</u> *In addition to the assumptions of Theorem 2, suppose that*

$$\varphi_k(\cdot,0) \leq \varphi_k(\cdot,T),\ \psi_k(\cdot,T) \leq \psi_k(\cdot,0) \text{ in } \Omega$$

for $k = 1,\ldots,n$. *Then problem* (1)-(2)' *admits a solution* u *which belongs to* $[W_q^{2,1}(Q)]^n$ *for any* $q \in [p,\infty[$.

6. For the case when (1) is replaced by

$$\begin{cases} u_k \leq \psi_k, \\ E_k u_k \leq F_k(u,\nabla u_k), \\ [E_k u_k - F_k(u,\nabla u_k)](u-\psi_k) = 0 \\ \text{in } Q \text{ for } k = 1,\ldots,n, \end{cases} \tag{1)'$$

or even by

$$E_k u_k = F_k(u,\nabla u_k) \quad \text{in} \quad Q \text{ for } k = 1,\ldots,n, \tag{1)''$$

we introduce the following terminology. We say that $\varphi \equiv (\varphi_1,\ldots,\varphi_n)$, with

φ_k as in Section 3, is a *lower solution (of* (1)" *with respect to* $\psi \equiv (\psi_1,\ldots,\psi_n)$, with ψ_k as in Section 3) if, for h = 1,...,m and k = 1,...,n,

$$E_k\varphi_{k,h} \leq F_k(v^{(k)},\nabla\varphi_{k,h}) \text{ in } Q$$

whenever $v^{(k)} = (v_1,\ldots,v_{k-1},\varphi_{k,h},v_{k+1},\ldots,v_n)$ with $v_j \in C^0(\overline{Q})$, $\varphi_j \leq v_j \leq \psi_j$ on $\overline{Q}$ for $j \neq k$. The definition of an *upper solution* ψ *(of* (1)" *with respect to* φ) is at this point obvious. Note that, except when n = 1, a solution $u \in [W_p^{2,1}(Q)]^n$ of (1)"-(2) whose components lay between φ and ψ need not be either a lower or an upper solution. Various types of conditions on the f_k's ensuring the existence of lower and upper solutions can be found, for instance, in [5].

The reason for introducing the above notions is provided by the following simple result, which reduces the investigation of both (1)' and (1)" to that of (1).

<u>PROPOSITION.</u> *Same assumptions about the functions* f_k,φ_k,ψ_k *as in Section 3. The set of all solutions to* (1) *coincides with the set of all solutions* u *to* (1)' *which in addition satisfy*

$$u_k \geq \varphi_k \text{ on } \overline{Q} \text{ for } k = 1,\ldots,n$$

if φ *is a lower solution, and even with the set of all solutions* u *to* (1)" *which in addition satisfy*

$$\varphi_k \leq u_k \leq \psi_k \text{ on } \overline{Q} \text{ for } k = 1,\ldots,n$$

if ψ *is in its turn an upper solution.*

REFERENCES

1. Aronszajn, N.: Le correspondant topologique de l'unicité dans la Théorie des équations différentielles, Ann. of Math. <u>43</u> (1942), 730-738.

2. Bebernes, J.W. and K. Schmitt: Invariant sets and the Hukuhara-Kneser property for systems of parabolic partial differential equations, Rocky Mountain J. Math. <u>7</u> (1967), 557-567.

3. Krasnosel'skii, M.A. and P.G. Sobolevskii: The structure of the set of solutions of equations of parabolic type, Dokl. Akad. Nauk SSSR <u>146</u> (1962), 26-29; Engl. Transl. in Soviet Math. Dokl. <u>3</u> (1962), 1230-1234.

4. Lasry, J.M. and G.M. Troianiello: On the set of solutions to a semilinear parabolic equation, Comm. Partial Differential Equations 7 (1982), 1001-1021.

5. Troianiello, G.M.: Bilateral constraints and invariant sets for semilinear parabolic systems, Indiana Univ. Math. J. 32 (1983), 563-577.

6. Troianiello, G.M.: Maximal and minimal solutions to a class of elliptic quasilinear problems, Proc. Amer. Math. Soc. 91 (1984), 95-101.

7. Troianiello, G.M.: Regular solutions to nonvariational obstacle problems for parabolic operators, Math. Nach. (to appear).

G.M. Troianiello
Dipartimento di Matematica
Università di Roma I
Italia

A VISINTIN

Partial differential equations with hysteresis

Hysteresis effects can appear in phase transitions; ferromagnetism, plasticity, supercooling and superheating are typical examples. Hysteresis appears also in filtration through porous media (in biology and chemistry, among others). The only systematic mathematical research on hysteresis seems to be that conducted by Krasnosel'skii, Pokrovskii and co-workers [1]. The present author has been (and is still) studying hysteresis especially in connection with partial differential equations [3], [4]; a review of results can be found in [5].

1. Continuous hysteresis functionals

We consider a one-dimensional ferromagnetic body and denote by u and w the magnetic intensity and the magnetic induction fields. In general, at each time t, w(t) depends on $u|_{]-\infty,t]}$, not just on u(t). We shall assume that $u \in C^0_c(]-\infty,T])$ (space of continuous functions with compact support). For instance, if u is as in Fig. 1, then (u,w) follows the path of Fig. 2. Thus, setting

$$u^t(\tau) := u(t+\tau) \qquad \forall \tau \in \mathbb{R}^-, \quad \forall t \in \mathbb{R}, \tag{1.1}$$

we have

$$w(t) = F(u^t) := [\tilde{F}(u)](t) \qquad \forall t \in \mathbb{R}. \tag{1.2}$$

Thus $F : C^0_c(\mathbb{R}^-) \to \mathbb{R}$ and $\tilde{F}$ is a <u>Volterra</u> (i.e. <u>casual operator</u>). In this paper we shall assume that

$$\tilde{F} : C^0_c(\mathbb{R}) \to C^0(\mathbb{R}), \quad \tilde{F} \text{ is continuous.} \tag{1.3}$$

F will be named <u>hysteresis functional</u> if

$$\begin{cases} \forall v \in C^0(\mathbb{R}^-), \ \ \forall s : \mathbb{R}^- \to \mathbb{R}^- \text{ surjective, continuous and} \\ \text{not-decreasing, } F(v) = F(v_o s) \ \ (\text{"\underline{Rate-independence}"}). \end{cases} \tag{1.4}$$

An example of hysteresis functional is sketched in Fig.3. Here the couple (u,w) can move horizontally in both directions, but on $\overline{AB}$ (respectively on $\overline{CD}$) w can just decrease (increase, respectively). Notice that in this case for any $t > 0$, $F(u^t)$ depends just on $F(u^o)$ and on $u|_{[0,t]}$.

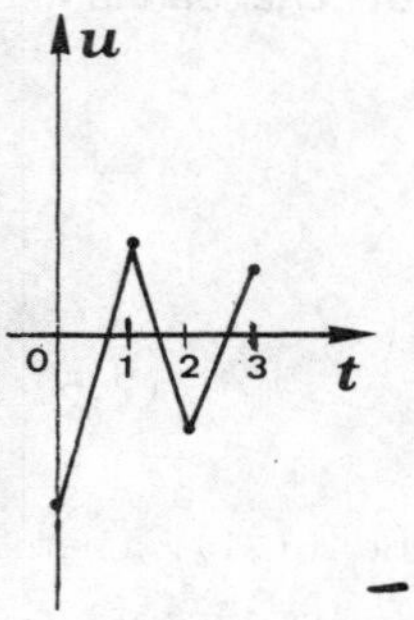

Fig. 1

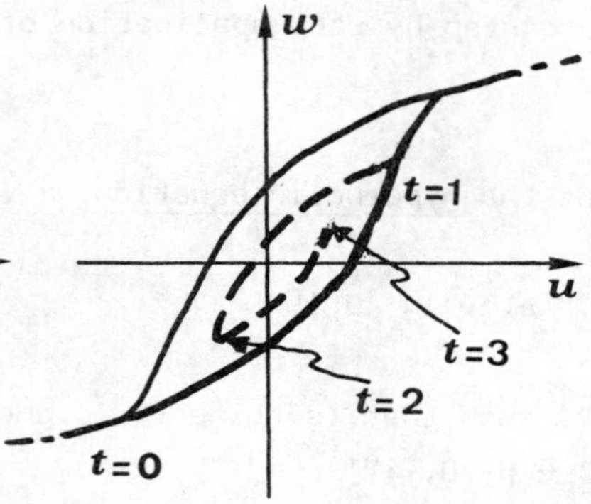

Fig. 2

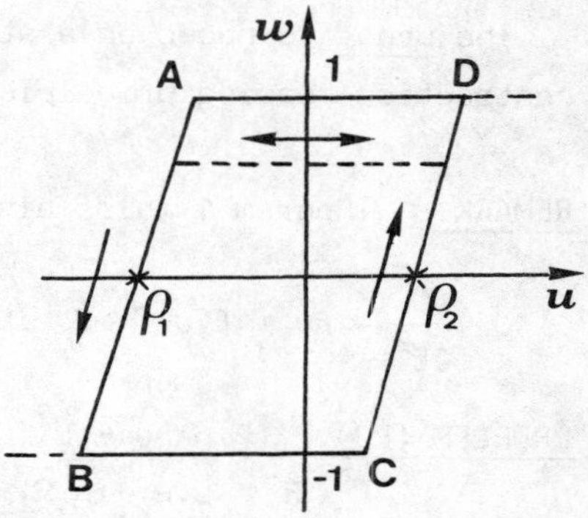

Fig. 3

2. P.D.E.'s with Volterra functionals

Henceforth F and $\tilde{F}$ will be as in (1.2). Let Ω be a bounded domain of $\mathbb{R}^N$ $(N \geq 1)$, $V \subset L^2(\Omega) = L^2(\Omega)' \subset V'$ be a Hilbert triplet, $A \in L(V,V')$ be elliptic and symmetric, $f \in L^2(\Omega \times]0,T[)+W^{1,1}(0,T;V')$ and

$$u^o \in L^1_{loc}(\Omega \times \mathbb{R}^-), \ u^o(\cdot,0) \in V, \ u^o(x,\cdot) \in C^0_c(\mathbb{R}^-) \ \text{ a.e. in } \ \Omega.$$

<u>PROBLEM</u> <u>(P1).</u> Find $u \in L^2(0,T;V) \cap H^1(0,T;V')$ such that defining

$$u(x,t) := \left\{ \begin{array}{ll} u^o(x,t) & \text{if } \ t \leq 0 \\ & \\ u(x,t) & \text{if } \ 0 < t \leq T \end{array} \right\} \text{ a.e. in } \Omega \tag{2.1}$$

and u^t as in (1.1), then $u^t(x,\cdot) \in C^0_c(\mathbb{R}^-)$ a.e. in Ω, $\tilde{F}(u) \in L^2(0,T;V')$ and

$$\frac{\partial u}{\partial t} + Au + \tilde{F}(u) = f \quad \text{in } V', \ \ \text{a.e. in }]0,T[. \tag{2.2}$$

THEOREM 1. *Assume that*

$$\mathcal{F} \textit{ is Lipschitz-continuous in } C_c^o(\mathbb{R}^-). \tag{2.3}$$

Then (P1) *has one and only one solution such that*

$$u, \tilde{\mathcal{F}}(u) \in H^1(0,T;L^2(\Omega)),\ u \in L^\infty(0,T;V). \tag{2.4}$$

The proof is based on a standard step by step application of the Banach contraction-mapping principle.

REMARK. Theorem 1 holds also for the hyperbolic equation

$$\frac{\partial^2 u}{\partial t^2} + Au + \tilde{\mathcal{F}}(u) = f \quad \text{in } V', \quad \text{a.e. in }]0,T[. \tag{2.5}$$

PROBLEM (P2). Find $u \in L^2(0,T;V)$ such that, setting (2.1) and (1.1), then $u^t(x,\cdot) \in C_c^o(\mathbb{R}^-)$ a.e. in Ω, $\tilde{\mathcal{F}}(u) \in H^1(0,T;V')$ and

$$\frac{\partial}{\partial t}\tilde{\mathcal{F}}(u) + Au = f \quad \text{in } V', \quad \text{a.e. in }]0,T[. \tag{2.6}$$

REMARK. If $N = 1$, $V = H_o^1(0,a)$ and $A = -\dfrac{\partial^2}{\partial x^2}$, then (2.6) corresponds to Maxwell equations with no displacement current for a one-dimensional ferromagnetic body.

THEOREM 2. *Assume that* (1.3) *holds and that*

$$\begin{cases} \exists\alpha > 0:\ \forall v \in C_c^o(\mathbb{R}^-),\ \forall t_1 < t_2,\ \textit{if } v \textit{ is linear in} \\ [t_1,t_2],\ \textit{then} \\ [\mathcal{F}(v^{t_2})-\mathcal{F}(v^{t_1})]\cdot[v(t_2)-v(t_1)] \geq \alpha[v(t_2)-v(t_1)]^2. \end{cases} \tag{2.7}$$

$$\exists C_1, C_2 \in \mathbb{R}^+:\ \forall v \in C_c^o(\mathbb{R}^-),\ |\mathcal{F}(v)| \leq C_1|v(0)| + C_2 \tag{2.8}$$

$$\textit{the injection } V \to L^2(\Omega) \textit{ is compact.} \tag{2.9}$$

Then (P2) *has at least one solution such that*

$$u \in H^1(0,T;L^2(\Omega)) \cap L^\infty(0,T;V),\ \tilde{F}(u) \in L^2(\Omega;C^0([0,T])). \tag{2.10}$$

Outline of the proof

(i) Approximation. Let $m \in \mathbb{N}$, $k = \frac{T}{m}$.

PROBLEM $(P2)_m$. Find $u_m^n \in V$ for $n = 1,\dots,m$ such that, setting

$$\hat{u}_m(x,t) := \begin{cases} u^o(x,t) & \text{in } \mathbb{R}^- \\ \text{linear interpolate of } \hat{u}_m(x,nk) := u_m^n(x), & \text{in }]0,T[, \end{cases} \tag{2.11}$$

$$\frac{F(\hat{u}_m^{nk}) - F(\hat{u}_m^{(n-1)k})}{k} + Au_m^n = \frac{1}{k}\int_{(n-1)k}^{nk} f(t)dt \quad \text{in } V',\ n=1,\dots,m. \tag{2.12}$$

At each step, $\hat{u}_m^{(n-1)k}$ is known, hence $F(\hat{u}_m^{nk})$ depends just on u_m^n: $F(\hat{u}_m^{nk}) := \varphi_n(u_m^n)$. By (2.7), $\varphi_n : \mathbb{R} \to \mathbb{R}$ is coercive. Then $(P2)_m$ has one (and only one) solution.

(ii) Estimates. We multiply (2.12) by $\frac{u_m^n - u_m^n}{k}$ and sum in n; by the coerciveness of φ_n, using a standard procedure we get

$$\|u_m\|_{H^1(0,T;L^2(\Omega)) \cap L^\infty(0,T;V)} \le \text{Constant}, \tag{2.13}$$

whence by (2.8)

$$\|u_m\|_{L^2(\Omega;L^\infty(0,T))} \le \text{Constant}. \tag{2.14}$$

(iii) Limit. Then, possibly extracting a subsequence,

$$u_m \to u \quad \text{weakly star in } H^1(0,T;L^2(\Omega)) \cap L^\infty(0,T;V) \tag{2.15}$$

$$\tilde{F}(u_m) \to w \quad \text{weakly star in } L^2(\Omega;L^\infty(0,T)). \tag{2.16}$$

By (2.9), for any $r \in]\frac{1}{2},1[$ also the injection $H^1(0,T;L^2(\Omega)) \cap L^\infty(0,T;V) \to$ $\to H^r(0,T;L^2(\Omega)) \subset L^2(\Omega;C^0([0,T]))$ is compact; hence, possibly extracting a further subsequence,

$$u_m(x,\cdot) \to u(x,\cdot) \quad \text{in } C^0([0,t]), \quad \text{a.e. in } \Omega; \tag{2.17}$$

then by (1.3)

$$\tilde{F}(u_m(x,\cdot)) \to \tilde{F}(u(x,\cdot)) \quad \text{in } C^0([0,t]), \quad \text{a.e. in } \Omega, \tag{2.18}$$

and by (2.16) $w = \tilde{F}(u)$. Taking $\varepsilon \to 0$ in (2.12) we get (2.6).

REMARKS.

(i) The uniqueness of the solution of (P2) is an open question. Assumption (2.7) does not allow to reproduce the standard argument. On the other hand, the monotonicity property

$$\int_0^T [F(v_1^t) - F(v_2^t)]\cdot[v_1(t) - v_2(t)]dt \geq 0 \quad \forall v_1, v_2 \in C^0(\mathbb{R}^-) \tag{2.19}$$

is too restrictive for hysteresis functionals.

(ii) Theorem 2 holds also for (2.2), (2.5) and for

$$\frac{\partial}{\partial t}\left[\frac{\partial u}{\partial t} + \tilde{F}(u)\right] + Au = f \quad \text{in } V', \quad \text{a.e. in }]0,T[; \tag{2.20}$$

existence holds also for the Cauchy problem governed by

$$\frac{\partial u}{\partial t} + a(x,t)\frac{\partial u}{\partial x} + \tilde{F}(u) = f \quad \text{a.e. in } \mathbb{R} \times]0,T[\,. \tag{2.21}$$

In all of these cases (2.7) is not necessary.

(iii) The previous results have been confirmed by numerical tests [2].

3. Discontinuous hysteresis functionals

Now we consider the case when F does not preserve the time continuity. The simplest example is given by the "relay functional" sketched in Fig. 4; there $\rho_1, \rho_2 \in \tilde{\mathbb{R}}$ $(:= \mathbb{R} \cup \{-\infty\} \cup \{+\infty\})$, $\rho_1 < \rho_2$. For any $t > 0$, $w(t)$ $(= \pm 1)$ is determined by $u|_{[0,t]}$ and by $w(0) = w^0 = \pm 1$, where $w^0 = -1$ (respectively $w^0 = 1$) if $u \leq \rho_1$ ($u \geq \rho_2$, respectively). Thus $w(t) = [F_{(\rho_1,\rho_2)}(u,w^0)](t)$.

Relationships of this sort appear in ferromagnetism, biology, chemistry.

$\rho_1 = -\infty$, $\rho_2 \in \mathbb{R}$ and $\rho_1 \in \mathbb{R}$, $\rho_2 = +\infty$ correspond to "complete irreversibility".

The relay functional $F_{(\rho_1,\rho_2)} : (u,w^0) \mapsto w$ is not closed with respect to the strong topology of $C^0([0,t])$ for u and the weak star topology of $L^\infty(0,T)$ for w. Its closure is a multivalued functional: $\overline{F}_{(\rho_1,\rho_2)}$; this is sketched in Fig. 5, which can also be obtained from Fig.3 letting the slope of $\overline{AB}$ and $\overline{CD}$ go to infinity.

Multivalued hysteresis functional of relay type. We first notice that the total variation of w in [0,t] is bounded; indeed w has a variation equal to 2 each time that u reaches one of the critical values ρ_1 and ρ_2, and the

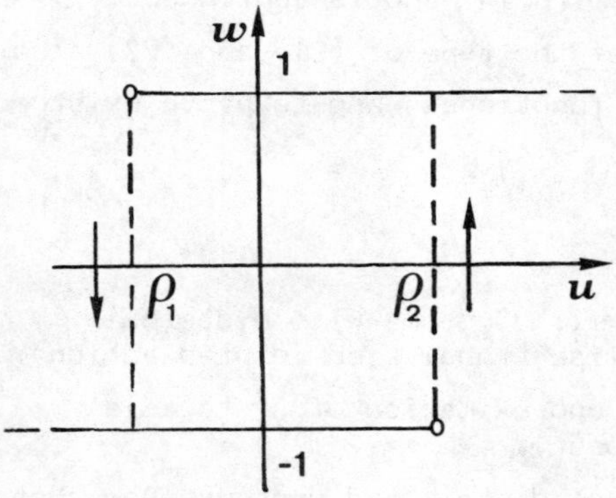

Fig. 4

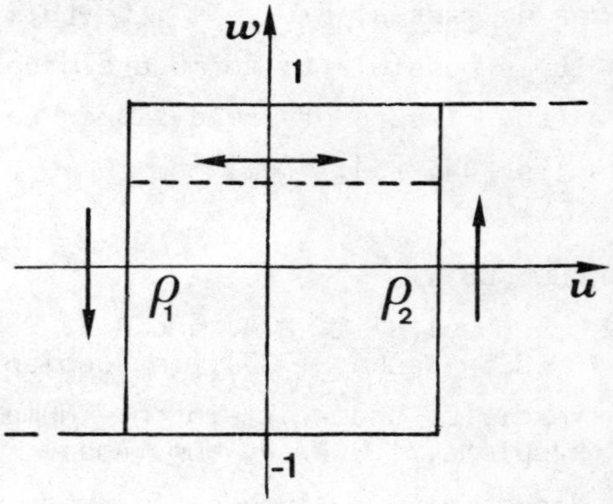

Fig. 5

uniformly continuous function u can have just a finite number of oscillations between ρ_1 and ρ_2. We also set

$$\alpha_\rho(v) := (v-\rho_2)^+ - (v-\rho_1)^- ;\ \beta_\rho(v) := v - \alpha_\rho(v)\ \forall v \in \mathbb{R},\ \forall \rho = (\rho_1,\rho_2).$$

Then $w \in \overline{F}_{(\rho_1,\rho_2)}(u,w^0)$ if and only if

$$w\cdot[\alpha_\rho(u)-v] \geq |\alpha_\rho(u)| - |v| \quad \forall v \in \mathbb{R},\ \text{in }]0,T[, \tag{2.22}$$

$$_{C^0(]0,t[)'}\langle w',\ \beta_\rho(u)-v\rangle_{C^0(]0,T[)} \geq 0 \quad \forall v \in C^0([0,T];[\rho_1,\rho_2]), \tag{2.23}$$

$$[w(0)-w^0]\cdot[\beta_\rho(u(0))-v] \geq 0 \quad \forall v \in [\rho_1,\rho_2]\ . \tag{2.24}$$

This formulation can be extended also to vector-valued functions $u,w: [0,T] \to \mathbb{R}^3$.

Preisach model - Composing several relays with different thresholds, a large class of (in general discontinuous) Volterra functionals can be obtained. Let μ be a finite measure in the (extended) "Preisach plane" $P := \{\rho = (\rho_1,\rho_2) \in \tilde{\mathbb{R}}^2 \,|\, \rho_1 \leq \rho_2\}$. We set

$$[F_\mu(u,\{w_\rho^o\}_{\rho\in P})](t) := \int_P [F_\rho(u,w_\rho^o)](t)d\mu_\rho \ , \tag{2.25}$$

for any compatible argument.

F_μ is rate-independent, as any F_ρ is so. If $\mu \geq 0$, F_μ fulfils (2.7). If μ has no masses, $F_\mu(\cdot,w^o) : C^o([0,T]) \to C^o([0,T])$ and is continuous.

It is possible to formulate problems of the type of (P1) and (P2) also for (the closure of) relays and Preisach functionals and to prove existence results [4], [5].

REFERENCES

1. Krasnosel'skii, M.A. and A.V. Pokrovskii: Systems with hysteresis, (Moscow Nauka, 1983) (in Russian; English translation in preparation).
2. Verdi, C. and A. Visintin: Numerical approximation of hysteresis problems, I.M.A. J. Num. Anal. (to appear).
3. Visintin, A.: A model for hysteresis of distributed systems, Ann. Mat. Pura Appl. 131 (1982), 203-231.
4. Visintin, A.: On the Preisach model for hysteresis, Nonlinear Anal. TMA 9 (1984), 977-996.
5. Visintin, A.: Partial differential equations with hysteresis functionals In "Proceedings of an I.N.R.I.A. symposium", Versailles, 1983 (North-Holland, 1984).

A. Visintin
Istituto di Analisi Numerica
Consiglio Nazionale delle Ricerche
C. so C. Alberto
27100 Pavia
Italia